LEARNING
MATHEMATICS

LEARNING MATHEMATICS

A program for classroom teachers

Ross McKeown

Heinemann

Portsmouth, NH

Heinemann Educational Books Inc.
70 Court Street, Portsmouth, NH 03801
Offices and agents throughout the world

© Ross McKeown 1985, 1990
Preface © 1990 by Lois McNulty

First published in 1985 by Robert Andersen and Associates,
1091 Glenhuntly Road, Glenhuntly 3163, Australia

ISBN 0-435-08304-X

Typeset in Univers by Bookset.
Printed in Australia by Globe Press Pty Ltd.

Contents

Acknowledgements

Sincere thanks to the following people for their contributions:

Judy Kelly for typing the original manuscript;

John Punshon for the photographs;

Marg Hookey who has permitted me to incorporate her use of vocabulary
   cards for the basic operations which are referred to in Curriculum
   Development Centre, Maths Shop Project, *Approach to Number*;

Bill Newton for his advice and permission to use the headings from the
   measurement guides;

the staff of Boronia Heights Primary School who have helped trial and
   refine many of the activities; and

Murray Landt whose unqualified support and encouragement made this
   book possible.

Preface

Teachers, like their students, learn best by doing. We get ideas from other teachers, from courses, workshops and sometimes even from textbooks. We teach, not "from the text," but from our own, constantly changing notebook of ideas, games, and materials, depending on the needs of our students.

Learning Mathematics, written by Ross McKeown, an Australian elementary school teacher and college lecturer, reads like a teacher's own notebook. It is the product of several years' work with elementary-level students.

The book is comprehensive and detailed, yet concise. It does not try to reinvent a math curriculum, but instead gives practical suggestions for making math learning more productive and meaningful.

McKeown uses a chart and outline format to present an activity-based math program for the elementary years. Central to his program is the belief that every child learns at his or her own rate and that learning should be continuous and ungraded. The sections are based on the idea that children do not have set work to be covered at particular grade levels, but rather progress through sections at their own rate for the first seven years at school. Children within the same grade level, therefore, will be working in more than one section.

McKeown gives suggestions for organizing the program, with sample schedules for making it work in a real classroom. Knowing that teachers will pick and choose, and finally use their own favorite games and activities, McKeown only gives examples and suggestions — for whole-group activities, small-group games, and independent work.

McKeown encourages teachers to evaluate students' progress in an ongoing way by observing them playing games, by asking them simple, direct questions and listening to their answers, or by having them perform a few calculations. At the end of each concept discussed, he gives specific suggestions for evaluating the child's understanding of it. A chapter on evaluating student progress includes sample tests for each level and record-keeping suggestions for the whole class, which stress continuity in each student's math learning program.

McKeown has chosen to organize elementary-level math concepts into seven "strands", and gives examples of games, teaching methods and activities to use at every age level. They are:
- Automatic Response (Number Facts)
- Counting — Pattern and Order
- Basic Operations
- Place Value
- Processes (like two-digit multiplication)
- Basic Properties (like equality, commutative law, etc.)
- Fractions.

This overall program chart for the first seven years is contained on two pages (pp. 2–3), making it easy for teachers at every level to use. Measurement is treated separately, but in the same manner, with an easy-to-scan chart (pp. 100–101) and suggested activities.

With McKeown's program, students "get friendly" with mathematical ideas by using materials, playing games, talking and experimenting with concepts before they are required to memorize facts and perform calculations. And this is exactly the kind of math learning recommended by the National Council of Teachers of Mathematics in its 1989 *Curriculum and Evaluation Standards for School Mathematics*, a significant document aimed at reforming math education in this country.

"The need for curricular reform . . . is clear," the Standards state. The present curriculum "fails to foster mathematical insight, reasoning, and problem solving and emphasizes rote activities . . . Children begin to lose their belief that learning mathematics is a sense-making experience."

The first four curriculum standards established by the NCTM for grades kindergarten through eight are: mathematics as reasoning; and mathematical connnections. Computation is eighth on the list. As for teaching methods, the Standards urge use of manipulative materials, cooperative work, discussion and questioning; it discourages teaching by telling, written practice, rote memorization. For middle school grades, the Standards urge instructional practices that will actively involve students in exploring, analyzing, and applying math in real-world contexts.

Most of the materials McKeown refers to are familiar to U.S. teachers: Multilinks, Unifix Cubes, Cuisenaire Rods. Some materials and terms, however, are unique to Australia. In order to retain the continuity of McKeown's text, we will explain those Australian terms and materials here:

Each "section" refers to approximate grade levels in U.S. schools — sections A would be kindergarten or preschool, section B would be first grade, etc.

Vulgar fractions — fractions expressing numbers less than 1, like ½ and ¼.

Beat the Tape — a commercial audiotape package that recites mental math problems.

Welford Blocks (p. 36) — small flat blocks of various geometric shapes, these are similar to Creative Publications' Pattern Blocks.

Infant Squares (p. 53) — flat plastic squares measuring 15 mm × 15 mm.

Wainwright Fraction Kit and S.A. fraction kit — commercially available in Australia, they are plastic pieces which students can use to show fractional relationships.

Icy-pole sticks — popsicle sticks.

M.A.B. (p. 9) — Multibased Arithmetic Blocks are similar to Dienes Blocks or Base 10 Blocks and are available through Creative Publications.

This is a book for those teachers who enjoy math — and for those who wish they did. It is a book teachers will keep, not on the shelf, but on their desks — open.

Lois McNulty
Cashman Elementary School
Amesbury, Massachusetts

Introduction

I Do and I Understand

It is suggested that we shift the emphasis from teaching to learning, from our experience to the children's, in fact from our world to their world.

Zoltan Dienes

In order to fully understand mathematical concepts it is essential that children experience them in concrete form. That is, in order to develop new concepts they must actively manipulate materials whilst verbalizing their ideas and understandings in their own everyday language.

Children learn best when they can:
1. learn through understanding;
2. use concrete materials to learn new concepts;
3. learn at their own rate;
4. work in small groups;
5. talk about their developing comprehension with other children and the teacher;
6. work with a teacher who realizes that children go through stages of cognitive development;
7. work with a teacher who *enjoys* mathematics and can transmit this enjoyment to the children;
8. see their mistakes as a step towards a better understanding;
9. play mathematical games to reinforce new concepts;
10. apply their mathematical knowledge in their day-to-day lives;
11. avoid dull, repetitious, 'busy' work.

This book aims to provide teachers with a mathematics program based on an active learning approach. In each grade, students are grouped according to their ability and the organizational requirements of the teacher. They commence Section A of the course in their first year of schooling and proceed through the program for seven years. It is anticipated that by the end of seven years most children will have at least mastered Section G.

The students work through nine measurement topics over seven years. The planned sequence of these follows grade levels, with each topic being divided into seven units of work.

PROGRAM

PROGRAM	SECTION A	SECTION B	SECTION C	SECTION D
AUTOMATIC RESPONSE (Number Facts)				
COUNTING – PATTERN AND ORDER	**EARLY NUMBER** • Vocabulary • Oral counting to 10 • Pattern • Recognize and write numerals to 10 • Sorting — classification • Ordering • Cardinal Number to 10 • Ordinal Number to 10th	• Forwards by 1, 5, 10 to 100, 2s to 20 • Backwards by 1s to 20 • One more/less than to 10 • Number before/after • Counting of groups to 10 • Patterns — matching, completing, creating • 1-to-1 correspondence • Ordinal to 20th • Cardinal to 20 • Limit to 10	• Maintain and strengthen	• Forwards by 1, 2, 5, to 100, 4s to 40 • Backwards by 1, 5, 1 from 100 • Patterns — matching completing and creating • Classify — one attribute • Count on in a series • Two more/less than 100 • Smallest to largest (and vice versa) • Group counting to 2 (groups to 10) • Ordinal to 20th • Doubling and halvin to 20

BASIC OPERATIONS		• Addition
		• Multiplication — 12
		• Inter-relationship of addition and multiplication
			to 12
			• Subtraction
			  — take away
			  — difference
			• Division
			  — quotition
			  — partition
			• Inter-relationship of operations
				to 24
				• Addition
				• Multiplication
				• Subtraction
				  — take away
				  — difference
				• Division
				  — quotition
				  — partition
				• Complementary addition
				• Inter-relationship of four operations
				• Solve, create and manipulate equation

PLACE VALUE		• Recognition of numbers to 20 • Write numbers to 10 • Recognize and write words to 10	• Maintain and strengthen	• Write numbers to 10 • Words to 20 • Recognize words to 1 • Place value to 99
PROCESSES				
BASIC PROPERTIES		• Equality and inequality	• Continue to develop • Utilize measurement	• Informal treatment o mathematical laws a axioms
FRACTIONS		• Informally treated	• Informally treated	• Informally treated

SECTION E	SECTION F	SECTION G	SECTION H
• Addition and subtraction facts to 20 • Multiplication facts 2, 3, 4, 5, 10 times tables	• 6, 7, 8, 9, 11, 12 times tables	• Maintain and strengthen	• Maintain and strengthen
• Counting to 1000 • Completing counting sequences • Serial addition and subtraction • Doubling and halving • Ordinal to 100th	• Counting to 10 000 to tenths • Serial addition, subtraction, multiplication and division • Doubling and halving • Completing counting sequences • Words to thousands	• Counting to 100 000 to ·01 • Doubling and halving — factors and fractions • Completing counting sequences • Serial addition, subtraction, multiplication and division • Rounding off to nearest 10, 100, 1000, $\frac{1}{10}$ $\frac{1}{100}$ etc.	• Counting to 1 000 000 to .001 • Completing counting sequences • Doubling and halving — fractions, multiplication, division
to 150 • Maintain and strengthen • Greater than/less than > <	• Solve • Manipulate • Real life → maths sentences • Maths sentences → real life	• Record physical situations in mathematical language and symbols in clear, unambiguous statements	• Maintain and strengthen • Verbalize in everyday language all mathematical sentences and concepts covered
• Write words to 100 • Write numerals to 1000 • Place value to 999 • Extended notation	• To 1000s • To 10 000s • To 100 000s • Extended notation • Tenths • Renaming	• To 100 000 • To ·01 • Extended notation • Renaming	• To 1 000 000 • To ·001 • Index notation • Other bases
• Addition — 2 addends — 3 digits • Multiplication — 2 digits × tables • Problem solving — addition, multiplication	• Addition — 3 addends — 4 digits • Subtraction, 3 digits • Multiplication — 3 digits × tables to 12 • Division — short (partition)	• Addition — 4 addends thousands • Subtraction — 4 digits • Short multiplication — 4 digits × tables to 12 • Long multiplication — 3 digits × 2 digits • Short division	• All processes treated at a degree of difficulty that challenges and extends the children • Averages
• Informal treatment of mathematical laws and axioms	• Informal treatment of mathematical laws and axioms	• Thorough understanding of mathematical laws and axioms	• Maintain and strengthen • Extensive use of pro-numerals to ensure thorough understanding
• Relationship between two numbers, e.g. 1/3 • As an operator $\frac{1}{2}$ of 8 • As a counting number	• Equivalent fractions	• Addition of $\frac{1}{3} + \frac{2}{3}$ $\frac{1}{3} + \frac{1}{6}$ • Subtraction of $\frac{3}{4} - \frac{1}{4}$ $\frac{1}{3} - \frac{1}{6}$ • Multiplication $3 \times \frac{1}{4}$ • Add decimals and subtract decimals (to hundredths) • Multiply decimals $3 \times ·4$ • Percentage as a fraction (concept only)	• Addition $\frac{1}{2} + \frac{1}{3}$, mixed numbers • Subtraction $\frac{1}{4} - \frac{1}{5}$, mixed numbers • Multiplication $\frac{1}{3} \times \frac{1}{4}$ • Division $2 \div \frac{1}{3}, \frac{1}{2} \div 2$ • Add and subtract decimals to thousandths • Multiplication — tenths and hundredths • Division of decimals $684.5 \div 5$ • Percentage as an operator • Probability

ORGANIZATION

An Approach for Developing Mathematical Concepts

Most of the activities are set out using the following concept development model.

Teachers who require further information about this model could view the video* 'Learning Maths' which was produced by Melbourne State College at Boronia Heights Primary School in 1983.

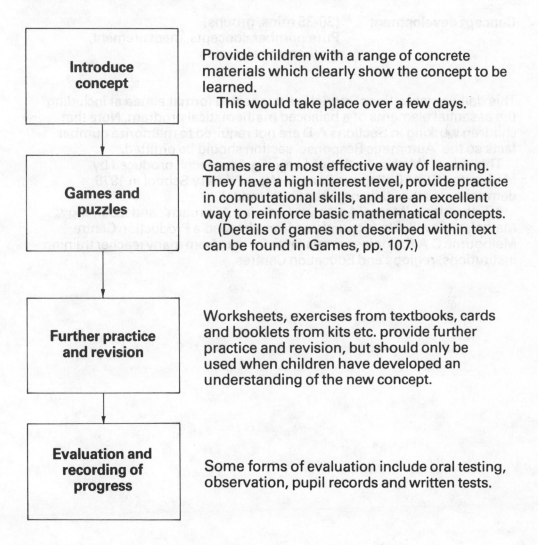

Introduce concept

Provide children with a range of concrete materials which clearly show the concept to be learned.
This would take place over a few days.

Games and puzzles

Games are a most effective way of learning. They have a high interest level, provide practice in computational skills, and are an excellent way to reinforce basic mathematical concepts.
(Details of games not described within text can be found in Games, pp. 107.)

Further practice and revision

Worksheets, exercises from textbooks, cards and booklets from kits etc. provide further practice and revision, but should only be used when children have developed an understanding of the new concept.

Evaluation and recording of progress

Some forms of evaluation include oral testing, observation, pupil records and written tests.

Daily Organization

Automatic response   (10 mins, whole grade activity)
Tables; addition and subtraction facts to 20.

**Counting — pattern
and order**   (10-15 mins, whole grade activity)

Concept development   (30-35 mins, groups)
Pure number concepts, measurement,
mathematical games.

This daily organizational model is a suggested format aimed at including the essential elements of a balanced mathematical program. Note that children working in Sections A-D are not required to memorize number facts so the 'Automatic Response' section should be omitted.

The video* 'Maths, An Active Learning Approach', produced by Melbourne State College at Boronia West Primary School in 1978, demonstrates this daily organizational model.

*Both videos, 'Maths, An Active Learning Approach', and 'Learning Maths', are available for purchase from the Media Production Centre, Melbourne C.A.E. Copies are available for loan from many teacher training institutions, regions and Education Centres.

Weekly Planning

There is no set formula that will suit all teachers because there are so many factors that affect organization. Considerations include:
1. number of children in the class;
2. range of abilities;
3. number of groups that can be organized effectively;
4. physical environment;
5. available resources.

However there are parts of the maths program that can be taken with the whole class (Automatic Response and Pattern and Order) before any grouping need occur.

When considering the number of groups to be organized, the following points are important:
1. If you divide an average-sized class into only two groups, they will be too big for most mathematical games, measurement activities and concept development with concrete materials. The range of abilities within the groups will also present problems.
2. Too many groups create organizational difficulties.
3. Teachers must feel in control of, and comfortable with, the chosen form of organization.

The following weekly plan is a suggested format that some teachers have found works well. It uses a Grade 5 lesson as a model.

Work program entry

Automatic Response (See section entitled 'Automatic Response' for directions on how to play these games.)
1. 'Tables Football'
2. 'Tables Baseball'
3. 'Tables Knock Out'
4. 'Number Grids'
5. 'Buzz'

Pattern and Order
1. Counting on beadframe — tenths
2. Individual number charts
3. Study 8's chart
4. Counting — Cuisenaire staircase
5. Serial multiplication and division

Measurement
Area: Use cm^2 grid paper to make rectangles of various areas.
Conservation of area activities.

Concept Development
pure number
(See section entitled 'Games and Puzzles' for directions on how to play games.)

	Concept	Game
*Group 1	Place value to hundredths	'Decicus'
Group 2	Multiplication of fractions $3 \times \frac{1}{5}$	'Vulcation'
Group 3	Short multiplication algorithm	'Roll a Product'
Group 4	Equivalent fractions	'Fracto'

*Children are grouped according to ability.

If three groups are organized then the work program has one less group and the weekly plan on the following page is also simplified.

The one constant in the weekly plan is 'measurement' and this can be planned for in advance.

In addition to this work, each of the groups will proceed through activities with concrete materials, games and activity sheets.

Note that the development of a new concept does not fit neatly into a five-day cycle, in fact some concepts might span two or three weeks.

WEEKLY PLAN

	Monday	Tuesday	Wednesday	Thursday	Friday
Automatic Response (10 mins) whole class	'Tables Football'	'Tables Baseball'	'Tables Knock Out'	'Number Grids'	'Buzz'
Pattern and Order (15 mins) whole class	Counting on beadframe — tenths	Individual number charts	Counting Cuisenaire staircase	Study 8's chart	Serial multiplication and division.
Concept Development (30-35 mins) Children work in ability groups					
Group 1	Measurement — area	Place value cards → ·01	Abacus	Game — 'Decicus'	Activity sheet
Group 2	Number line Multiplication of fractions $3 \times \frac{1}{5}$	Measurement — area	S.A. 64 fraction kit	Game — 'Vulcation'	Activity sheet
Group 3	M.A.B. 10 short multiplication algorithm	M.A.B. 10	Measurement area	M.A.B. 10	Game — 'Roll a Product'
Group 4	Equivalent fractions — paper folding	Fraction wall	S.A. 64 fraction kit	Measurement — area	Game — 'Fracto'

Preparing Activity Cards and Worksheets

When children use concrete materials to help develop mathematical concepts, the teacher's role is more demanding, subtler, and far more rewarding than traditional teaching.

A group of children who are beginning to learn a new concept will use a range of materials over a few days, although only some of these activities will require the teacher to work with the group. In this situation the teacher directs the use of materials, encourages discussion and generally guides the children through the activity. However there are many occasions when the teacher prepares an activity card or worksheet which directs the children in the use of the materials. Apart from a brief introduction, these sessions should require very little direct input by the teacher.

POINTS TO CONSIDER WHEN PREPARING ACTIVITY CARDS:

- The directions should be clear and concise.
- How the materials are to be used should be obvious.
- Only one concept should be developed per card.
- Attractive presentation motivates children to emulate the teacher's model.
- Illustrations often help clarify the activity.
- Pre-folded manilla folder card is an ideal size for a group activity card and becomes a valuable resource.

The following activity cards and sheets are examples of activities that children have worked through independently with success.
1. Place value — Section D: Tens and units
   Materials — Place-value cards
2. Addition of fractions — Section G
   Materials — number line sheets, chinagraph pencils and acetate boards
3. Area — Grade 5 (see Measurement Chart on p.00)
   Materials — centimetre grid paper, square tiles
4. Perimeter — Grade 4 (see Measurement Chart on p.00)
   Materials — Multilink, centimetre graph paper

Tens and Units

There are ten **digits** we can use to make numbers. They are:

0 1 2 3 4 5 6 7 8 9

Make these numbers with your **place value cards** and then **extend** them (take them apart), e.g.

| 2 | 6 | = | 2 | 0 | + | 6 |

45   =   40 + 5

63   =

86   =

19   =

24   =

Put these numbers together with your cards, e.g.

| 3 | 0 | + | 2 | = | 3 | 2 |

60 + 9   =

4 + 30   =

70 + 7   =

3 + 90   =

40 + 4   =

2 + 20   =

10 + 9   =

2 + 10   =

Adding Fractions on a Number Line

On the number line sheet there are nine number lines divided into different fractional parts.

Choose the number line for each problem and show the answer on the number line.

Example: $\frac{4}{8} + \frac{5}{8} = \frac{9}{8}$ or $1\frac{1}{8}$

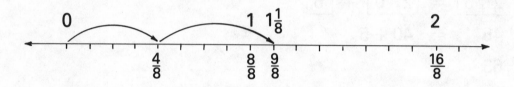

Do these on your number lines and write in the answers.

$\frac{1}{2} + \frac{2}{2}$ = $\frac{1}{7} + \frac{4}{7}$ =

$\frac{1}{3} + \frac{1}{3}$ = $\frac{3}{8} + \frac{2}{8}$ =

$\frac{2}{4} + \frac{3}{4}$ = $\frac{8}{9} + \frac{2}{9}$ =

$\frac{3}{5} + \frac{3}{5}$ = $\frac{2}{10} + \frac{4}{10}$ =

$\frac{4}{6} + \frac{5}{6}$ = $\frac{6}{10} + \frac{6}{10}$ =

Have you found out anything about adding fractions?

Make up some additions of fractions equations on the number lines.

Number Lines

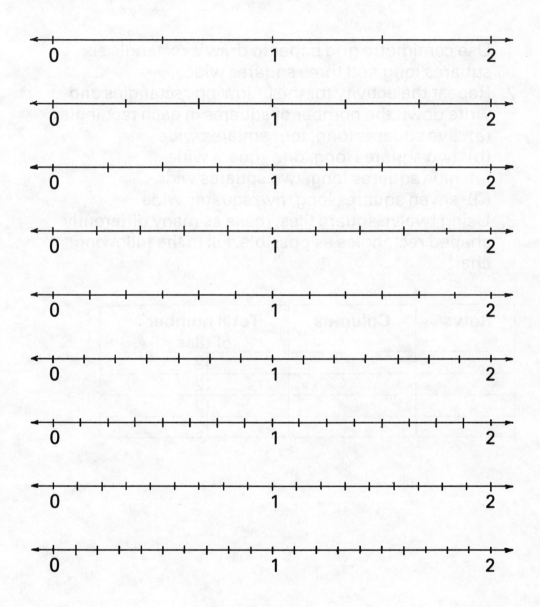

Area

1. Use centimetre grid paper to draw a rectangle six squares long and three squares wide.
2. Repeat the activity for the following rectangles and write down the number of squares in each rectangle:
   (a) five squares long, four squares wide
   (b) two squares long, one square wide
   (c) nine squares long, two squares wide
   (d) seven squares long, five squares wide
3. Using twelve square tiles, make as many differently shaped rectangles as possible. Fill in the following chart.

Rows	Columns	Total number of tiles
		12
		12
		12
		12

Perimeter

1. Use Multilink cubes to make shapes like this and then complete the following table.
   N.B. Each cube is 2 cm long.

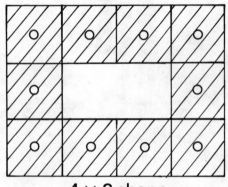

4 × 3 shape

Shape	Outside perimeter	Inside perimeter
4 × 3	28 cm	12 cm
3 × 3		
4 × 5		
6 × 3		
7 × 4		

2. On the centimetre graph paper, make shapes with the following perimeters:
   40 cm, 34 cm, 16 cm, 24 cm
3. On another sheet of graph paper draw:
   1. a square — perimeter 20 cm
   2. a rectangle — perimeter 18 cm
4. Make a shape of your own with Multilink and ask your neighbour to work out its perimeter.

AUTOMATIC RESPONSE

Automatic response, which includes addition and subtraction facts to 20 and multiplication tables to the 12-times table, starts at Section E and is considered an important aspect of the mathematics course. Children who have immediate recall of these number facts are able to complete calculations quickly and accurately, and may concentrate fully on new concepts to be learned. Children who do not have automatic response become frustrated and lose interest when calculations become drawn out and tedious.

The daily organizational plan suggests ten minutes of automatic response at the beginning of each mathematics session. If the games and activities chosen are sufficiently interesting and varied, this time becomes a popular part of each session.

The following ideas are all proven classroom activities and should provide sufficient variety and motivation to sustain children's interest and enthusiasm. These games are only a starting point and teachers will be able to add many more activities of their own.

'TABLES FOOTBALL'

The game is played between two teams of five. The two teams line up against each other in the key positions of Australian Rules Football (it can be modified for soccer or rugby).

Full Forward	Centre Half-forward	Centre	Centre Half-back	Full Back
X	X	X	X	X
X	X	X	X	X
Full Back	Centre Half-back	Centre	Centre Half-forward	Full Forward

The teacher starts the game by asking the two centre players a number fact. Whichever centre player answers first is allowed to 'send' the ball one place towards his or her team's goal. The teacher then repeats this operation between centre half-forward and centre half-back. A goal is scored when the full forward beats the full back and the ball is returned to the centre.

'TABLES BASEBALL'

Divide the class into four mixed-ability teams and give each team a batting order. Mark four bases across the width of the chalkboard.

Each team will require a small chalkboard, chalk and a duster.

The game starts with the four No. 1 batters beginning on first base. They make a straight line in front of first base facing the back of the room. The teacher asks a number fact and children have three seconds to record their answers. They then hold up their chalkboards and show the rest of the teams. Children who answer correctly move to second base. Children who answer incorrectly return to their team and are replaced by the next batter who must start at first base. Children who reach fourth base score a 'home run'. The team in front after ten minutes wins.

'TABLES KNOCK OUT'

Groups should be as close as possible in ability.

A group of children line up across the front of the room. The teacher asks a number fact. The first child to answer takes a step back and does not answer until the second round. The last child left at the front is eliminated. The remaining children step forward and the second round commences. This procedure continues until all but one child is eliminated. This child is the winner.

'BUZZ'

A group of children stand at the front of the room. The teacher chooses a number, e.g. 3. Children in turn start counting from 1 but must say 'buzz' if their number has a 3 in it (e.g. 13) or is a multiple of 3 (e.g. 6).

Children are out if they make a mistake or cannot answer in three seconds. The game continues until one child is left.

The difficulty can be increased by using two numbers, e.g. numbers 3 and 5 (1 2 buzz 4 buzz buzz 7 8 buzz buzz 11 etc.)

'NUMBER CIRCLE'

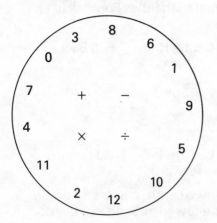

A number circle is drawn on the chalkboard. The teacher points to different numbers and signs, e.g. $6 \times 12 + 4 =$

The teacher selects a child to give answer.

'NUMBER GRIDS'

+	3	7	10...
4	7	11	
6			
5			

−	3	7	2
5	2		
4			
1			

×	8	7	12
3	24		
9			
2			

N.B. With the subtraction exercise, it must be the 'difference between' aspect.

'NOMOGRAM'

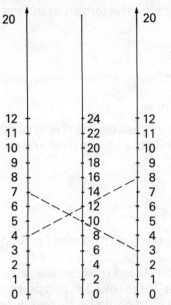

Children should use a ruler or some string to line up two numbers on the outside columns. The number in the centre that the line runs through is the sum of the two numbers, e.g.
4 + 8 = 12, 3 + 7 = 10.
Variation   The centre number minus one outside number equals the remaining number, e.g.
12 − 8 = 4, 10 − 3 = 7.

By adjusting the ruler slightly each time children begin to see a useful pattern, e.g. 1 + 9 = 10, 2 + 8 = 10, 3 + 7 = 10, 4 + 6 = 10 etc.

'MANIPULATING EQUATIONS'

The children are directed to write as many equations as possible using three numbers, e.g. 3, 7, 10 (3 + 7 = 10, 10 − 7 = 3) or 6, 3, 18 (3 × 6 = 18, 18 ÷ 6 = 3).

'THE FACTOR GAME'

1	2	3	4	5
6	7	8	9	10
11	12	13	14	15
16	17	18	19	20
21	22	23	24	25
26	27	28	29	30
31	32	33	34	35
36	37			

Prepare this grid on an overhead transparency with permanent markers. Divide the class into two teams. Each child should choose a number in turn. This number is a score for their team, e.g. a child may pick 30. The other team then automatically scores any number that is a factor of 30, i.e. 1, 2, 15, 6, 5, 3, 10 = total of 42.

As numbers are used they are crossed out with a water soluble marker until no numbers remain. Progressive totals are kept for each team.

37 would be a good number to start with because, being a prime number, the other team only gets 1 as it is the only other factor.

'SNAKE EYES' 1 to 10

Number cards 1 to 10 and draw a snake stretching the length of the ten cards on the reverse side, e.g.

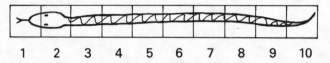

| 1 | 2 | 3 | 4 | 5 | 6 | 7 | 8 | 9 | 10 |

The game starts with the ten cards facing the class on the chalkboard ledge. A child or the teacher rolls 2 dice. The child then turns over any name for the sum of the two dice, e.g.

= 7

The child can turn over the number 7 card or the 6 and 1, 2 and 5 or 4 and 3 cards.

The game continues until all the cards are reversed to show the whole snake, or no move can be made, e.g. if another 7 was rolled at this point the game would have to end as neither the 4 or 9 card could be turned over.

The remaining numbers are added, e.g. $4 + 9 = 13$. The lower the final score the better the result. Groups can play against each other with the lowest score being the winner.

'CARDS' (traditional pack)

Children work in pairs, dealing out all the cards between them. The cards are turned face down in a pile in front of each child.

Court cards are worth 10 points each, Ace cards are worth one, other cards have face value.

Children turn over their top cards simultaneously. The first child to say the sum of the two cards takes the cards and puts them to the bottom of his or her pile. In the case of a draw, the cards are left in the centre and the child who has the next correct answer takes all these cards and adds them to her or his pile. The first child to collect all the cards wins.

Variations:
1. Multiplying the two cards.
2. Subtracting the value of the lowest card from the highest.
3. Dividing the highest card by the lowest card (for upper school children).

This can also be played with 10-sided dice.

'BINGO'

Prepare player cards. The teacher calls out number facts and the children cover the answers with counters. The first child to complete a row in any direction wins.

'FACTOR TREES'

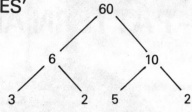

'BEAT THE TAPE'

Commercially available.

'SPEED TEST'

Provide each child with a duplicated sheet with number facts and give them a set amount of time in which to complete as many as possible.

'TABLES RACE'

Divide the class into groups and let one group play at a time. Ask them tables questions. The first to answer correctly takes one step forward and the first child to reach a given spot wins.

'MAGIC SQUARES'

6	1	8
7	5	3
2	9	4

1	15	14	4
12	6	7	9
8	10	11	5
13	3	2	16

17	24	1	8	15
23	5	7	14	16
4	6	13	20	22
10	12	19	21	3
11	18	25	2	9

Provide children with sufficient numbers to complete the magic squares.

COUNTING — PATTERN AND ORDER

Pattern is the underlying theme of mathematics. The skill of recognizing and using patterns is a valuable problem-solving tool for a child to learn to use for it can have a profound effect on the development of a child's mathematical understanding.

Pattern and order, of which counting is an important part, is best developed on a regular basis. Teachers who spend 10 or 15 minutes each day giving children a wide range of experiences in this area are providing an excellent basis upon which children can develop sound mathematical understandings.

The activities that follow are by no means an exhaustive list. They are only meant to demonstrate the variety of activities and materials that would be useful to use in this section of the course. Teachers should choose activities which are appropriate to their children's stage of progress.

COUNTING Pattern and Order

Provide children with:
1. variety in activities and materials (discrete* and structured)
2. time to discover.

Structural aids could include number charts, number boards, number lines, bead frames, Cuisenaire rods, abaci, Dienes M.A.B.

1. GROUP COUNTING

Will probably emerge from rhythmic counting e.g. clapping or stamping on every second or third number:

1 ② 3 ④ 5 ⑥ 7 ⑧ 9 ⑩

1 2 ③ 4 5 ⑥ 7 8 ⑨ 10 11 ⑫

2. CHALKBOARD RECORDING

This can lead to simple number charts.

by 2s		by 3s			by 4s		
1	2	1	2	3	1	2	3	4
3	4	4	5	6	5	6	7	8
5	6	7	8	9	9	10	11	12
7	8	10	11	12	13	14	15	16
9	10							
11	12							

Multiplication tables can be developed from this.

3. FINAL DIGIT PATTERNS

Let children work out final digit patterns when counting by different numbers (starting from zero), e.g.

by 2s 2 4 6 8 0
   4s 4 8 2 6 0
   6s 6 2 8 4 0
   8s 8 6 4 2 0

Ask the children to list five things they notice about the patterns which develop.
Try it with odd numbers.

---

*Schools abound with articles of interest — balls, skittles, bean bags, counters, flowers, cars in car park, chairs, trees, rows of boys and girls etc.

Patterns can appear in places other than the units, e.g.

8	9
16	18
24	27
32	36
40	45
48	54
56	63
64	72
72	81
80	90
88	99
96	108
104	
112	
120	
128	

Ask the children what they notice when counting by 9s.
Discovery should be encouraged, not forced.

4. CUISENAIRE RODS

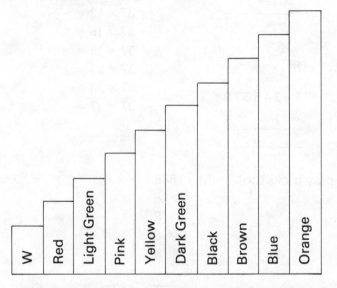

The teacher can make a staircase with jumbo magnetic Cuisenaire rods on the chalkboard, or each child can make their own staircase with normal-sized Cuisenaire rods.

The teacher or a child gives any rod a value then the class count together from white through to orange, e.g. the teacher may say 'Black equals 14'. The class will then count out the values of each rod when black equals 14 — 2, 4, 6, 8, 10, 12, 14, 16, 18, 20.

Teacher: 'Brown equals 2.' Class: '$\frac{1}{4}$, $\frac{2}{4}$' etc.

Teacher: 'Orange equals 1,000,000.' Class: '100,000, 200,000' etc.

Teacher: 'Orange equals ·1.' Class: '·01, ·02, ·03' etc.

5. THE ABACUS

Counting on an abacus reinforces place value ideas.
Children must exchange:
(a) one ten for 10 units
(b) one hundred for 10 tens
(c) one hundredth for 10 thousandths etc.

6. PATTERNS WITH NUMBERS

There are many mathematical curiosities that fascinate children. *Figuring — The Joy of Numbers* by Shakuntala Devi (Andre Deutsch, London, 1977) is an excellent resource book for ideas.

The following are just a few children could try. (Calculators could be used for more time-consuming calculations.)

(a)

$$1 \times 1 = 1$$
$$11 \times 11 = 121$$
$$111 \times 111 = 12321$$
$$1111 \times 1111 =$$
$$11111 \times 11111 =$$
$$111111 \times 111111 =$$

(b) Here is another oddity associated with 2.

```
 123456789
 123456789
 987654321
 987654321
         2
─────────
─────────
```

(c)

$$9 \times 9 + 7 =$$
$$98 \times 9 + 6 =$$
$$987 \times 9 + 5 =$$
$$9876 \times 9 + 4 =$$
$$98765 \times 9 + 3 =$$
$$987654 \times 9 + 2 =$$
$$9876543 \times 9 + 1 =$$

(d)

```
 12345679
      × 8
─────────
─────────
```

(e)

$$37 \times 3 =$$
$$37 \times 6 =$$
$$37 \times 9 =$$
$$37 \times 12 =$$
$$37 \times 15 =$$
$$37 \times 18 =$$
$$37 \times 21 =$$
$$37 \times 24 =$$
$$37 \times 27 =$$

(f)

```
 123456789
       × 9
──────────
──────────
```

(g) Prime numbers can play tricks too!

$$1^3 + 5^3 + 3^3 =$$

(h)

```
888
 88
  8
  8
  8
───
───
```

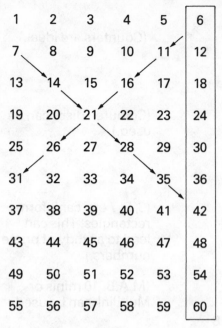

Study this 6s chart. What do you notice?

What happens when you count by 4s, 8s and 11s?

Do these things happen on other number charts?

7. COUNTING ON A NUMBER LINE

(a) Make a number line on the floor with masking tape. Young children can step on the number line as they count (forwards and backwards).

(b) Draw a number line on the chalkboard with a mixture of sugar and water. The markings will stay until washed off with water.

Different numbers can be placed on the intervals appropriate to the counting range e.g. decimals, vulgar fractions, whole numbers.

Children can be asked to complete the counting sequence.

(c) Draw number lines on duplicated sheets under acetate boards and use for a variety of activities.

8. BEAD FRAME

These can be used for counting in all sections of the course.

Count by 1s, 2s, 5s etc.

If one row equals 1, we can count by tenths. If the bead frame equals 1, we count by hundredths. Explore all counting possibilities, e.g.

$\frac{1}{100}$ $\frac{2}{100}$ $\frac{3}{100}$ etc.

·01 ·02 ·03 etc.

1% 2% 3% etc.

9. PROPERTY OF NUMBER

(a) Square numbers

(Counters are ideal.)

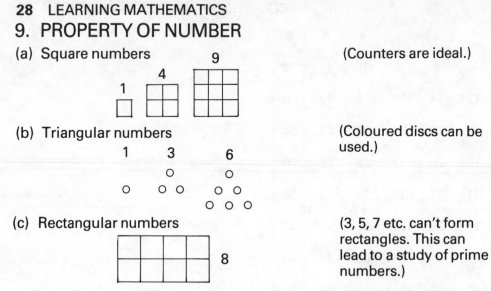

(b) Triangular numbers

(Coloured discs can be used.)

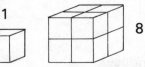

(c) Rectangular numbers

(3, 5, 7 etc. can't form rectangles. This can lead to a study of prime numbers.)

(d) Cubic numbers

(M.A.B. 10 minis or Multilink can be used.)

10. SERIAL ADDITION, SUBTRACTION, MULTIPLICATION AND DIVISION

$7 + 5 = 12$	$16 - 7 = 9$	$3 \times 4 = 12$	$16 \div 8 = 2$
$17 + 5 = 22$	$26 - 7 = 19$	$3 \times 40 = 120$	$160 \div 8 = 20$
$27 + 5 = 32$	$36 - 7 = 29$	$30 \times 40 = 1{,}200$	$160 \div 80 = 2$
$57 + 5 = 62$	$86 - 17 = 69$	$30 \times 400 = 12{,}000$	$1{,}600 \div 8 = 200$

It is important to do a lot of this work in Grades 3 to 6.

11. DOUBLING AND HALVING

3	64	1	$8 \times 4 = 32$	$12 \div 6 = 2$
6	32	$\frac{1}{2}$	$16 \times 2 = 32$	$24 \div 12 = 2$
12	16	$\frac{1}{4}$	$32 \times 1 = 32$	$48 \div 24 = 2$
24	8	$\frac{1}{8}$	$64 \times \frac{1}{2} = 32$	$96 \div 48 = 2$
48	4	$\frac{1}{16}$	$128 \times \frac{1}{4} = 32$	

12. ARRAYS

How many ways can you arrange 24 dots/counters as rectangles?

$6 \times 4 \qquad 4 \times 6$
$8 \times 3 \qquad 3 \times 8$
$2 \times 12 \qquad 12 \times 2$

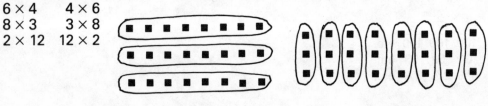

3×8 $\qquad\qquad\qquad\qquad$ 8×3

13. INTER-RELATIONSHIP OF BASIC OPERATIONS

Let children discover the patterns of the inter-relationship of the basic operations, e.g. use 5, 4, and 20 to make number sentences.

$5 \times 4 = 20$	$20 \div 5 = 4$	$\frac{20}{5} = 4$	$\frac{1}{4}$ of $20 = 5$
$4 \times 5 = 20$	$20 \div 4 = 5$	$\frac{20}{4} = 5$	$\frac{1}{5}$ of $20 = 4$
$20 = 5 \times 4$	$4 = 20 \div 5$	$4 = \frac{20}{5}$	$5 = \frac{1}{4}$ of 20
$20 = 4 \times 5$	$5 = 20 \div 4$	$5 = \frac{20}{4}$	$4 = \frac{1}{5}$ of 20

Use 6, 4, and 10

$6 + 4 = 10$	$10 - 6 = 4$
$4 + 6 = 10$	$10 - 4 = 6$
$10 = 4 + 6$	$4 = 10 - 6$
$10 = 6 + 4$	$6 = 10 - 4$

Important discussion can centre on why $+'^n$ and $-'^n$ could not be used with 5, 4 and 20, and $\times'^n$ and $\div'^n$ using 6, 4 and 10.

14. NUMBER SENTENCES

Ask children what number sentences they can see in the dot pattern, e.g.

$3 + 3 + 3 + 3 = 12$

$4 \times 3 = 12$

$12 - (2 \times 3) = 3 + 3$

$12 \div 3 = 4$
(how many)

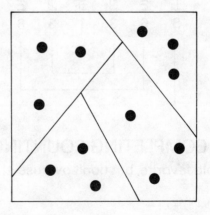

$12 \div 4 = 3$
(shared equally between four)
$\frac{1}{4}$ of $12 = 3$

$\frac{2}{4}$ of $12 = 6$

$\frac{3}{4}$ of $12 = 9$

$\frac{4}{4}$ of $12 = 12$

15. FINAL DIGIT PATTERNS

A most enjoyable and worthwhile counting activity.
   Let children explore the full range of final digit patterns e.g. 2, 3, 4 etc.

3s pattern

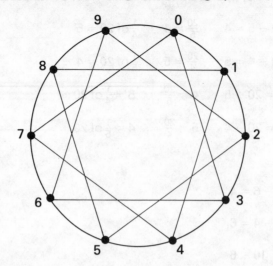

16. THE PATTERN OF OUR PLACE-VALUE SYSTEM

Notice how the unit, and not the decimal point, is the centre of our place-value system.
   Only when you focus on the units place, does the place-value system gain its balance and pattern.

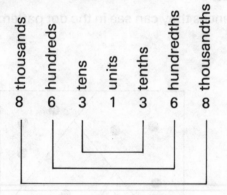

17. COMPLETING COUNTING SEQUENCES

(An old favorite, but don't overuse it!)

2, 4, 6, _ , _ , _ .

1, 7, 2, 14, _ , _ , _ .

18. EQUIVALENT FRACTION HUNT

1	1	2	3	4	5	6	7	8	9	10
2	2	4	6	8	10	12	14	16	18	20
3	3	6	9	12	15	18	21	24	27	30
4										
5										
6										
7	7	14	21	28	35	42	49	56	63	70
8										
9										
10										

During a pattern and order session, children could complete this chart as a counting activity.

The finished product provides an excellent session to reinforce children's understanding of equivalent fractions.

Any two rows can be compared to show equivalent fractions, e.g.:

(a) Rows 1 and 2 — $\frac{1}{2} = \frac{2}{4} = \frac{3}{6}$ etc.

(b) Rows 1 and 3 — $\frac{1}{3} = \frac{2}{6} = \frac{3}{9}$ etc.

(c) Rows 3 and 7 — $\frac{3}{7} = \frac{6}{14} = \frac{9}{21}$ etc.

19. NAPIER'S RODS

Children enjoy making their own sets of rods and this activity makes an excellent counting exercise. If the rods are made from cardboard, they are easy to manipulate and will last a long time. The rods are made by counting by the number at the top of each column, each figure being written in a separate box. An example is given for the '5' rod (see Figure A).

The rods are then cut out and used for short and long multiplication calculations.

A simple example follows showing how the rods can be used for long multiplication 153×26. The child puts together the 1, 5, and 3 rods with the index card (see Figure B).

The value of each place is added diagonally as shown.

First calculate 153×6, then calculate 153×2 and add a zero, i.e.

$$\begin{array}{r} 153 \\ \times\ 26 \\ \hline \end{array}$$

$$\begin{array}{rl} 918 & (153 \times 6) \\ +\ 3{,}060 & (153 \times 2 \text{ and add } 0) \\ \hline =\ 3{,}978 & \end{array}$$

Napier's Rods should only be used when children understand the short and long multiplication algorithms. They then become an interesting idea that can add greater depth to these concepts, and a novel way of correcting multiplication problems using the traditional algorithm.

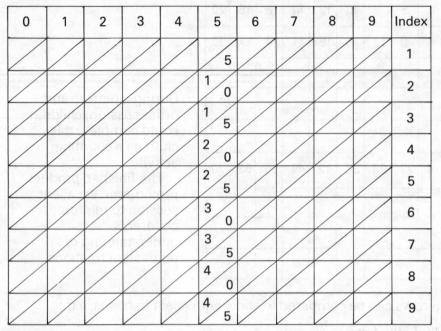

Figure A

Figure B

6 × 153

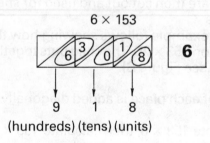

9   1   8
(hundreds) (tens) (units)

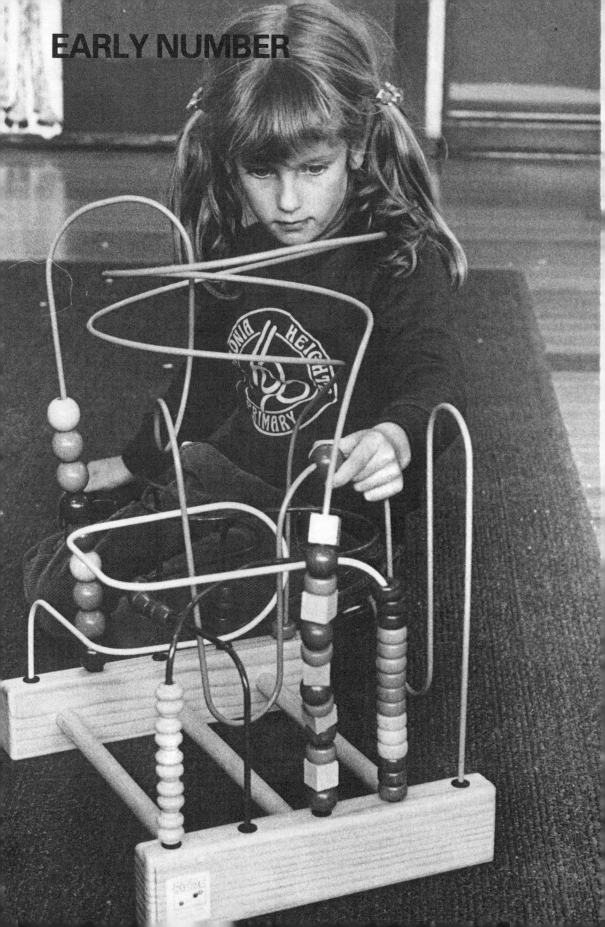

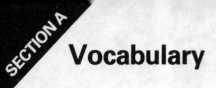

Vocabulary

The emphasis in this section is on the underlying concepts of vocabulary. Activities focus on concrete experiences and promote oral interpretation.

OBJECTIVES

The following vocabulary lists are included as a guide for the teacher. These words, and their underlying concepts, will form the basis of the teaching arising from free-play situations using concrete materials.

   This list is not meant to confine nor dictate. Teachers should be aware of their class's needs for realistic goals and further challenges.

SUGGESTED VOCABULARY

The following are terms which would be commonly used throughout the everyday life of a child at this stage of his or her development.

Functional
partner/line up/stop/go/front/back/twos/circles/next to/away from/in/inside/out/outside/corner/end-to-end/side-by-side/left/right/find out/start from/middle/begin/beginning

Time
now/morning/today/night/day/afternoon/after/yesterday/before/past/slowly/quickly/lunch/time/early/late/clock/lunchtime/playtime/in a little while/it's a long time

Colour
red/blue/yellow/green/orange/black/white/pink/grey/brown/purple/light/dark

Comparative
big/little, small/large, high/low, long/tall/short, fat/thin, thick/thin, top/bottom, biggest/smallest, same as/different from, behind/in front of/beside

Quantity
groups/pile/set/heap/money/buy/some/more/less/same/as much/not as much/fill/full/not the same/empty/many/few/not as full/overflowing/each/spill/number/balance

Shape
circle/square/rectangle/curve/straight/round/ring/triangle/flat/hexagon

Books
Opposites: 'Top and Bottom', 'Big and Little', 'Front and Back', 'Old and New', 'Fast and Slow'. Macdonald Educational, London, 1976

Oral Counting to 10

Most young children can say by rote the words from one through ten. This is essential for work with cardinal and ordinal number, even though little meaning may be associated with the words.

In addition to the traditional songs and rhymes, the following picture story books in the school library are an excellent resource to help develop children's oral counting and cardinal number.

Anno, Mitsumasa. *Anno's Counting Book*. Bodley Head, London, 1974

Carle, Eric. *1, 2, 3 to the Zoo*. Hamilton, London, 1969

Corbet, Ruby. *Animals One to Ten*. Golden Press, Sydney, 1979

Craig, Helen. *A Number of Mice*. Aurum, London, 1978

Dugan, Michael. *Nonsense Numbers*. Nelson, Melbourne, 1980

Elkin, Benjamin. *Six Foolish Fishermen*. Old Steam Wireless Factory

Fowler, Richard. *The Bean 1 2 3*. Longman, Harlow, 1981

Gretz, Susanna. *Teddybears 1 to 10*. Collins, London, 1973

Hargreaves, Rogers. *Count Worm*. Hodder & Stoughton, Leicester, 1976

——. *My Mr Men Counting Book*. Thurman, London, 1978

Hoban, Russell. *Ten What? A Mystery Counting Book*. Cape, London, 1974

Martin, Bill. *Ten Little Caterpillars*. Holt, Rinehart & Winston, New York, 1967

Niland, Deborah. *Birds on a Bough: A counting book*. Hodder & Stoughton, Sydney, 1975

Oxenbury, Helen. *Numbers of things*. Heinemann, London, 1967

Pavey, Peter. *One Dragon's Dream*. Nelson, Melbourne, 1978

Peake, Merle. *Roll Over*. Houghton Mifflin/Clarion, New York 1981

Peppe, Rodney. *Ten Little Bad Boys*. Penguin, Harmondsworth, 1978

Roffey, Maureen. *Farming with Numbers*. Bodley Head, London, 1972

Time for a Number Rhyme, Nelson, Melbourne, 1983

Trinca, Rod. *One Woolly Wombat*. Omnibus, Adelaide, 1982

Wadsworth, Olive. *Over in the Meadow*. Hamilton, London, 1973

Wildsmith, Brian. *1 2 3*. O.U.P., London, 1965

Counting Books. *This Old Man*.

CONCEPT Pattern

CONCRETE MATERIALS

Children should gain experience in:
1. matching patterns;
2. extending patterns;
3. creating patterns;
4. identifying missing elements in a pattern.

Useful materials which the children can use to gain this experience could include:

- pattern blocks — extremely popular with children
- Unifix — can include the beginnings of simple counting sequences
- Multilink — patterns can be three-dimensional
- farm animals — 1 horse, 2 cows, 1 horse etc.
- 3-D solids — cube, cylinder, cube, etc.
- coloured shapes pasted on paper which the children can either match or extend
- non-linear patterns

 Coloured shapes are pre-cut. Children choose shapes and paste them on to paper.

etc.

- dot patterns

 etc.

- Geoboards — allow for matching, extending and creating. Provide multi-coloured rubber bands for an extra variable.
- Figura — matching (commercially available)
- Cuisenaire rods

red yellow

- peg boards and pegs — the 20cm square plastic board (100 hole) is ideal. Children can create delightful patterns.
- Welford Blocks — children enjoy tessellating these shapes to form all-over patterns. Pattern blocks are more visually interesting.

CONCEPT Write the Numerals 0 to 10

CONCRETE MATERIALS

1. Ask the children to make numerals with their hands.
2. Children can practise writing the numerals by:
   (a) tracing over numerals already written
   (b) writing the numerals by copying.
3. The teacher can draw the numerals on large cards (30cm × 45cm). The first part of the numeral could be drawn in red; the second part could be blue.

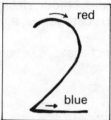

4. The children can make numerals with dough and plasticine etc.
5. Provide the children with salt trays and they can make imprints of numerals in them.
6. Geoboards and rubber bands can also be used by the children to make numerals.

GAMES

A child faces the chalkboard in front of the class while another child draws a numeral on their back. Children must press firmly and make the numeral large. The child being 'drawn on' must guess what numeral has been written.

FURTHER PRACTICE

1. Numeral cards can be displayed around the room or used to label objects, e.g. 'We have 4 goldfish.'
2. The teacher can cut numerals out of sandpaper and paste them on to cards. The children must close their eyes, feel the numeral and name it out aloud.
3. The children could be assisted to complete the numerals on number line templates (see p. 50 *Mathematics Their Way*).

EVALUATION

Observation over a period of time is the only real indication of a child's ability to correctly write the numerals 0-10. When a child consistently forms the numerals correctly this can be recorded on your mathematics progress chart, (see Evaluation, p. 114).

CONCEPT Sorting — classification

CONCRETE MATERIALS

1. 'Attribute blocks', 'Logic People', and 'Sea Sorts' can be used to develop this concept.
2. Various objects such as rocks, buttons, plastic counters, plastic animals etc. can be given to the children to classify.
3. Pictures of people, food, cars, aeroplanes etc. collected from magazines and newspapers can be classified into many different categories.
4. Wallpaper samples could be classified by texture, colour, design etc.
5. The children could even be 'sorted' into various categories, e.g. sex, colour of hair, style of dress, size etc.

These exercises can be integrated with other subject areas, e.g. science — classifying leaves, animals etc.

GAMES

1. 'What's in the Square?' Commercially available.
2. 'Same/Different' game. Make a partitioned sorting box. Design several cards so that some have differences and some are the same. The children should sort them into the various categories either you or they specify.

FURTHER PRACTICE

1. 'Sorting' can be done using the overhead projector and projecting geometric shapes. The teacher defines the categories and the children respond.
2. Provide the children with a box full of 'junk'.
   They can:
   (a) sort the contents by their own criteria;
   (b) sort by a given criteria;
   (c) determine how objects have been sorted;
   (d) pick out objects that do not belong to a set type of classification;
   (e) add objects to sorted groups.

EVALUATION

In order to develop logically and mathematically, a child must learn to classify. Classification is an ongoing process that ranges from a 'sorting' exercise in its earliest form to classification of rational, real and complex numbers in later years.

The above expectations, listed under Further Practice, would make a useful checklist at the end of a child's first year of mathematics.

CONCEPT Ordering

CONCRETE MATERIALS

1. Various objects can be ordered according to their different properties, e.g. straws can be ordered by length, measuring spoons by capacity, and cylinders by height or diameter.

    When working with the children, introduce quantitative terms such as longest, shortest, most, least, largest, smallest. Refer to the vocabulary list.

2. With the children's help, form a staircase with Cuisenaire rods. Ask the children questions such as 'Which rod is the shortest/longest?' 'Point to a rod longer than brown' etc.

3. Children can arrange sets of pictures to show a sequence.

4. Draw pictures on a stencil from a story. Read the story to the class. The children can cut out and order the pictures according to the events in the story.

GAMES

'What's Out of Order?' A set of objects is placed in order. The children close their eyes whilst one child changes two objects. The other children must then return the objects to the correct order.

FURTHER PRACTICE

Experience in seriating objects helps a child order numbers in a meaningful way, and ordering events provides readiness for measurement of time.

Guide children to discover the transitive property by posing the following: 'If A is shorter than B and B is shorter than C, then A must be shorter than C'. The transitive property is important in later work with comparisons of numbers.

EVALUATION

Children should progress through two stages:

1. Put objects in order by matching;

2. Order objects according to a given criteria, e.g.:
    (a) lengths from shortest to tallest;
    (b) objects from biggest to smallest;
    (c) colours from darkest to lightest;
    (d) sandpaper from roughest to smoothest;
    (e) sounds from lowest to highest;
    (f) towers from tallest to shortest.

CONCEPT Cardinal Number

CONCRETE MATERIALS

1. Objects that are the same (cards, marbles, straws, buttons) can be counted.
2. The children can rearrange a group of objects (conservation of number).
3. Children can be asked to swap an object in a group and to make a group of different objects.
4. Using a variety of materials, children can make a number of groups of the same size.
5. 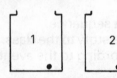 Labelled jars can be handed out which the children can fill with the appropriate number of objects.
6. Emphasize a number for a period of time by:
   (a) drawing it in coloured inks and pinning it around the room;
   (b) looking at pictures of the same number of objects (e.g. if the chosen number is 3, show children pictures of traffic lights, cricket stumps etc.)
7. Children can thread the given number of beads on to string.
8. Arrange a set of paper cups labelled from 0-10. Ask the children to put in the correct number of straws.
9. Cut out pictures showing a given number. The children can make a class mural with them.

GAMES

1. 'The Circle Game'
2. 'The Piggy Bank Game'
3. 'Spill the Beans'
4. 'People Counting Games'
5. 'The Pendulum Game'
   (Details of these games can be found in Mary Baratta-Lorton. *Mathematics Their Way*. Addison-Wesley, California, 1976.)

FURTHER PRACTICE

1. Cardinal number picture cards should be prominently displayed around the classroom. These can be made into mobiles and hung around the room.

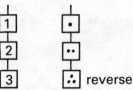

2. Children can practice on Unifix counting ladders.
3. Use every opportunity to count objects within the school environment, including the children themselves.
4. The teacher plays some music. When it stops, the children have to form groups of the number the teacher calls out. The children left over are eliminated.

STORIES

Stories which deal with cardinal number can be read aloud to the children.

Anno's Counting Book, Teddybears 1 to 10, One Unicorn ('The Child's World', Elgin, 1977), Ten Little Bad Boys, The Very Hungry Caterpillar (Penguin, Harmondsworth, 1974)

EVALUATION

In early experiences, a child counting to find the number of elements is desirable, but he or she should quickly identify the cardinal number of a set having up to five or six elements.

A child looking at five objects must understand that this 'fiveness' is an abstract property of the group and that it is independent of the kind of objects or how they may be arranged.

CONCEPT Ordinal Number

CONCRETE MATERIALS

1. Use a wide range of discrete materials, e.g. blocks, farm/zoo animals, counters, beads, chairs, felt board material etc.
2. Use the children themselves wherever possible — in line, queuing in games, taking turns, collecting materials.
   Constantly use and emphasize the appropriate vocabulary: after, before, between, in front of, last, next, left, right, middle, end, start, beginning, first, second etc.
3. Mobiles either made or bought can be used and liven up a classroom.
4. Attribute blocks. Ask the children questions such as 'What shape is the 1st?', 'What colour is the 4th?' etc.

GAMES

1. 'The Aeroplane Game'
2. 'Mischievous Kitten'
   Details of these games can be found in *Math Activities For Child Involvement* by Dumas and Schminke, Allyn & Bacon, Boston, 1977.

FURTHER PRACTICE

1. Children must understand that ordinal number requires:
   (a) a starting point;
   (b) a direction.
   To indicate the starting point, mark the first object to show direction, e.g. put a cross on the 3rd object

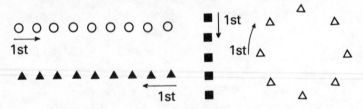

2. Draw three things in the second box, six things in the fourth box etc.

STORIES

Read stories in class such as:
12 days of Christmas
The Five Chinese Brothers
The Three Bears
The Three Little Pigs

EVALUATION

Observation of a child's work over a period of time is the best indication of his or her thorough understanding of this concept.

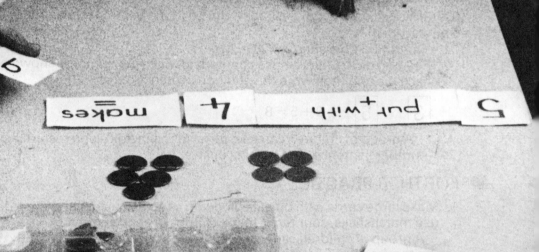

CONCEPT Basic Operation of Addition

CONCRETE MATERIALS

This concept is developed through three stages, over a period of time.

oral/concrete _____ vocabulary cards _____ maths symbols

1. A child puts a group of three objects with a group of four objects and then counts the objects to determine the number altogether. This activity is teacher-directed.

2.

| 3 | put with | 4 | makes | 7 |

The teacher adds signs to the cards.

| 3 | put with $+$ | 4 | makes $=$ | 7 |

The teacher discusses the signs with the children and they begin to see the signs are a short way of writing the words.

$3 + 4 = 7$

GAMES

(Details of all games not described can be found in the section entitled Games, pp.107).

1. 'Operation Charades'
2. Roll two dice. Children use counters to make the numbers shown on each die and then write the addition sentence, e.g.

| 3 | | 5 | $3 + 5 = 8$

A group of children can take turns to roll the dice and record the number sentence. The highest answer wins.

FURTHER PRACTICE

1. Make large cards with objects glued on as activities for children, e.g. use matchsticks, counters, Unifix, stamps of cent coins etc.

A group of children can work through this activity independently of the teacher. Provide children with the material that is on the card.

2. Let the children use a variety of materials and then make number sentences to record the physical arrangements, e.g. counters, farm animals, Unifix, bottle tops etc.
3. Make a class-size number line (3-4 metres long) with masking tape on the floor. Children then 'step out' number sentences, e.g. take four steps, take three more (4 + 3 = 7).
4. Mathematical balance.

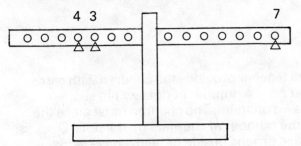

3 and 4 makes 7

Let children make and record addition sentences of their own.
Children must realize that a state of equality is reached when the apparatus balances (liken it to a see-saw).

EVALUATION

1. Ask children to complete addition number sentences using materials, e.g. 6 + 4 = ☐

2. In class make up 'real life' stories from addition sentences, e.g. 4 + 5 = 9 — 'I had four marbles. I bought five more so I had nine marbles altogether.'
   Reverse this by giving the children a 'story' and asking them to tell you what the addition sentence would be.

Note: number sentences of this kind 5 + ☐ = 8, should not be given to evaluate a child's understanding of addition. This is solved by using complementary addition which is an aspect of subtraction.

CONCEPT Basic Operation of Multiplication

CONCRETE MATERIALS

Stages of development

1.
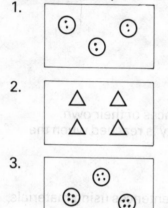
The teacher provides the children with cards that have a number of dots within shapes drawn on them. The children must place the same number of counters on the dots. Discuss. Make groups beside or below base cards.

2.
Place counters in groups on base cards, e.g. 'Put two counters in each shape on your card.'

3.
Use vocabulary cards to explain what is on the base card.

3 lots of	4

4. Put out groups to match cards

lots of 2	5	lots of 2

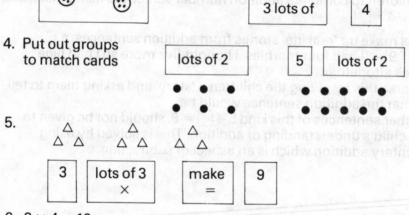

5.

3	lots of 3 ✕	make =	9

6. 3 × 4 = 12
Use maths symbols only.

GAMES

Play some lively music. When the teacher stops it, he or she calls out a number. The children must form groups of that number, with children leftover being eliminated. Discuss formations towards the end of the game. 'There are four groups of three left. How many children are left?'

FURTHER PRACTICE

1. Put down two lids and place three buttons on each lid. Ask the children how many buttons there are altogether.
2. As above with plastic farm animals and paddocks.
3. Use mixed objects on lids.

4. Take three strings and place four plastic beads on each. Ask the children to say how many beads there are altogether. Check by counting.
5. Make a tower of five blue Unifix cubes, and then make a tower of five red cubes. Ask the children how many cubes there are altogether. Discuss the answer.

EVALUATION

1. Evaluation will occur through oral discussion and observation.
2. Ask the children to complete number sentences using materials, e.g.
   $4 \times 2 = \square$ $3 \times 4 = \square$
3. Make up 'real life' stories in class from multiplication sentences and vice versa, e.g. $5 \times 2 = 10$ — 'There were five paddocks with two sheep in each paddock, making ten sheep altogether.'
4. Do not confuse rote memorization of some multiplication facts with an understanding of the basic operation of multiplication. Children must be able to physically represent multiplication number sentences.

CONCEPT Basic Operation of Subtraction — take-away

CONCRETE MATERIALS

The 'take-away' aspect of subtraction involves the physical removal of objects from a larger group. This can be taught in three stages, over a period of time.

oral/concrete _____ vocabulary cards _____ maths symbols

1. A child takes two objects from a group of six.

2. | 6 | take away – | 2 | leaves + | or | makes + | 4 |

3. $6 - 2 = 4$
5. Activity cards using a range of discrete materials can be used.
6. 'Take-away' can be related to counting back on a number line.
7. Put out 10 counters. Roll a die and take away the number of counters shown on the die. Ask the children to guess how many are left. Check by counting out aloud.

GAMES

1. 'Operation Charades'
2. Skittles • • • • Children roll a ball at skittles and then
   • • •   write a 'take away' number sentence
   • •   to describe what happened, e.g. if 4
   •   were knocked over, $10 - 4 = 6$.

FURTHER PRACTICE

1. Apply the same game for counters and die, using Unifix and die instead.
2. Use groups of plastic zoo animals and farm animals in paddocks for further practice.
3. Different or similar shapes on a felt board or magnetized buttons on a chalkboard (right-hand section) can also be used.
4. The children can step back on a number line on the floor, e.g. 'Start on eight, step back three and land on five, $(8 - 3 = 5)$

EVALUATION

1. Ask children to complete number sentences using materials, e.g. $8 - 5 =$
2. Ask children to make up 'real life' stories from subtraction sentences and vice versa, e.g. Seven birds were sitting on the fence. Three flew away leaving only four birds.
3. Children should be able to write a number sentence for a pictorial representation of the 'take-away' aspect of subtraction, e.g.

        $8 - 2 = 6$

CONCEPT Basic Operation of Subtraction — difference

CONCRETE MATERIALS

1. This aspect involves the child in comparing two groups of objects and determining how many more the larger group has, thus giving a difference.

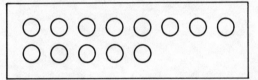

The difference between six and four is two.

$6 - 4 = 2$

2. Use any discrete materials e.g. counters, animals, Unifix, shells etc.
3. One-to-one matching base cards (use self-adhesive shapes, drawings, rubber stamps etc.) can also be used.

○ ○ ○ ○ ○ ○ ○ ○
○ ○ ○ ○ ○

Children cover the shapes on the card with counters and read the card to the teacher.

Later children can complete the 'difference between' card, e.g.

The difference between 8 and 5 is

Don't write the sum out using mathematical symbols (i.e. $8 - 5 = 3$) yet. This will come at a later stage.

GAMES

1. 'Operation Charades' $6 - 3 = 3$ (difference)
2. 'Scrabble' letters. Put your name on the card, one letter per square. Put a friend's name underneath, e.g.

Are they different? Which has more letters? Which has less? How many letters different are they?

FURTHER PRACTICE

1. Put ten lions and two elephants on a table or the floor. Ask the children to guess how many more lions there are. Check the estimates.
2. Put a handful of beads and a handful of counters on the floor. Ask the children how many more beads there are than counters (or vice versa). Get the children to check their answers.
3. Help the children make towers with Unifix or Multilink. Compare and discuss the sizes of the towers.

4. Mixed bag. Ask children to match a group of five things with a group of three things.
5. Give the children a set of cards with numbers on them and a pile of counters. Ask each child to take a card and count out a pile of counters adding up to that number. Then ask them to do the same with another card. What is the difference between the two groups?

EVALUATION

Evaluation will occur through oral discussion and observation. Eventually when a child reads a number sentence, he or she should be able to interpret it in two ways, e.g. $(7 - 4 = 3)$ 7 take away 4 equals 3 or the difference between 7 and 4 is 3.

CONCEPT Basic Operation of Subtraction — complementary addition

When children have a thorough understanding of the 'take away' and 'difference' aspects of subtraction, complementary addition is introduced.

This course deliberately leaves a space between the teaching of the 'take away' aspect, the 'difference' aspect and complementary addition. Experience has shown that three different physical manipulations for a subtraction number sentence can be confusing for many young children. E.g. 7 − 4 = 3:

The 'take away' aspect is taught in Section C.

7 take away 3 leaves 4.

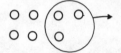

The 'difference' aspect is also taught in Section C.

The difference between 7 and 4 is 3.

○ ○ ○ ○ (○ ○ ○)
○ ○ ○ ○

Complementary addition is taught when all other operations are thoroughly understood.

What must I add to 4 to make 7?

○ ○ ○ ○ ○ ○ ○
○ ○ ○ ○ (● ● ●)

Complementary addition is the way most shopkeepers give change and many children are familiar with this practice.

If a child buys something for 4c and gives the shopkeeper 20c, the change is usually given in the following way:

4c	plus	1c	makes	5c
	plus	5c	makes	10c
	plus	10c	makes	20c

The change is 16c because 4c + 16c = 20c

Many adults, when asked to solve a subtraction problem, mentally use complementary addition without realizing what name is given to the method they used, e.g.

$$104 - 39$$
$$39 + 1 = 40$$
$$+ 60 = 100$$
$$+ 4 = 104$$
$$104 - 39 = 65$$

Children who become proficient at this method can often use it instead of a pencil and paper algorithm.

$$
\begin{array}{r}
1,0\overset{9}{\cancel{0}}\overset{9}{\cancel{0}}0 \\
-\ \ 864 \\
\hline
136
\end{array}
\qquad
\begin{array}{r}
1,000 \\
-\ ,864 \\
\hline
136
\end{array}
$$

$$864 + 36 = 900$$
$$+\ 100 = 1,000$$
$$\overline{136}$$

Decomposition   Equal additions   Complementary
                                                addition
                                             (done mentally)

Encourage children to have the flexibility to examine subtraction problems and decide whether complementary addition might be more appropriate than the algorithm they have been taught.

CONCEPT Basic Operation of Division — how many (quotition)

CONCRETE MATERIALS

Stage 1
Oral/concrete, e.g. put your six animals in two's. How many lots of two have you got?

Stage 2

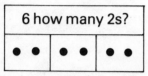

Use sticks or straws to separate the groups.

Stage 3

8	how many 2s ÷	equals =	4

Stage 4
$10 \div 5 = 2$

GAMES

Many of the activities below could form the basis of games that groups or the class could play.

FURTHER PRACTICE

Direct the children through the following activities:

1. Count out eight shapes. Put them in groups of two. How many two's did you make?
2. Using plastic fish and infant squares, count out ten fish. Put five in each pond. How many ponds are full?
3. On prepared activity sheets, circle objects to make groups of two. How many groups did you make?
4. Using plastic farm animals and fences, make some paddocks. Count out twelve sheep. Put three in each paddock. How many paddocks do you need?
5. Collect a handful of shells. Put them in rows of five. How many shells have you? How many rows did you make? How many are left over?

EVALUATION

The terms 'divided by' and 'divided into' have no place at this stage, for they do not inform the child of which real life arrangement she or he is to undertake and describe.

When a child has learnt the quotition and partition aspects of division, he or she should be able to say for the following number sentence $10 \div 2 = 5$: '10 shared between 2 equals 5' and when asked 'In 10, how many groups of two are there?', she or he should be able to answer '5'.

CONCEPT Basic Operation of Division — sharing (partition)

CONCRETE MATERIALS

Stage 1 — Oral/concrete
Children share objects fairly in a variety of ways, e.g. put the same amount of sheep in each pen.

Stage 2

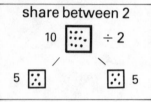

Stage 3

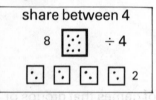

8 share between 4 = 2
÷

Stage 4

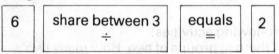

Stage 5
$10 \div 5 = 2$

GAMES

Give each group of children a number of tennis balls (or any other objects) that they can divide equally amongst their group. Tell them to share the balls with their group. The first group to share fairly wins.

FURTHER PRACTICE

1. Using plastic strings and beads, ask the children to share the beads fairly on each string.
2. Children can share acorns into egg cartons.
3. Show children a number of marbles and ask them to guess fair shares. They can check to see if they were right.
4. Find opportunities to let children 'share out' classroom materials. Discuss results.
5. Set up situations where fair shares cannot be given. Ask the children to share seven marbles between them. Discuss the problems with the class.

EVALUATION

Sharing is one skill that children acquire at an early age. They will use many differing strategies to share. The terms 'fair shares', 'equal shares' and 'same' are often used to explain a partition activity.

THE PHYSICAL OPERATION OF DIVISION

Children should see that division is a **re-arranging** operation. They are not altering the original size of the group, but manipulating and describing the groups formed.

CONCEPT Basic Operations and Their Inter-relationship

CONCRETE MATERIALS

1. Arrange a group of discrete materials in many ways. Record the patterns using basic operations, e.g.

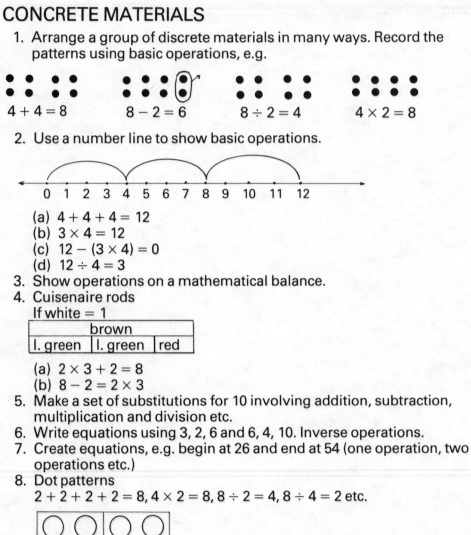

$4 + 4 = 8$ $8 - 2 = 6$ $8 \div 2 = 4$ $4 \times 2 = 8$

2. Use a number line to show basic operations.

    (a) $4 + 4 + 4 = 12$
    (b) $3 \times 4 = 12$
    (c) $12 - (3 \times 4) = 0$
    (d) $12 \div 4 = 3$
3. Show operations on a mathematical balance.
4. Cuisenaire rods
   If white = 1

brown		
l. green	l. green	red

   (a) $2 \times 3 + 2 = 8$
   (b) $8 - 2 = 2 \times 3$
5. Make a set of substitutions for 10 involving addition, subtraction, multiplication and division etc.
6. Write equations using 3, 2, 6 and 6, 4, 10. Inverse operations.
7. Create equations, e.g. begin at 26 and end at 54 (one operation, two operations etc.)
8. Dot patterns
   $2 + 2 + 2 + 2 = 8, 4 \times 2 = 8, 8 \div 2 = 4, 8 \div 4 = 2$ etc.

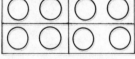

GAMES

1. 'Operation Charades'
2. 'Roll on Equation'

FURTHER PRACTICE

1. Children should solve, create and manipulate number sentences.
2. Help children to write 'real-life' situations in number sentences and then give them number sentences from which to construct 'real-life' situations.
3. Like language, we need to 'punctuate' to ensure our number sentences say what we mean. Encourage children to use brackets to avoid ambiguities, e.g. $3 + (4 \times 3) =$
4. Avoid order of operations conventions at all cost.

EVALUATION

Remember:

1. Mathematical sentences (e.g. $7 + 2 = 9$) are simply concrete situations that a child has observed or thought of, and recorded.
2. Mathematical sentences provide a form of communication.
3. Equations are only one form of mathematical sentence, e.g. $3 + 6 \neq 10$ and $6 > 4$ are also number sentences.
4. Materials should be available so that physical manipulation produces mathematical situations.

PLACE VALUE

CONCEPT Place Value to 999

CONCRETE MATERIALS

1. Using icy-pole sticks, children can make bundles of ten with rubber bands, and then put ten bundles of ten together to make bundles of a hundred.
2. Place-value cards can be used, e.g.

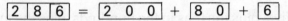

$$\boxed{2\ 8\ 6} = \boxed{2\ 0\ 0} + \boxed{8\ 0} + \boxed{6}$$

3. M.A.B. 10 clearly shows the value of numbers.
4. The abacus is the most abstract of the materials, as a disc on a spike takes its value depending on its place.

GAMES

1. 'Build a Number'. The digits 0 to 9 are written on cards and three cards are drawn randomly. Children place them in order to make largest or smallest number.
2. 'Make a Flat' — same idea with icy-pole sticks and abacus.
3. 'Wipe Out' — calculator activity.

FURTHER PRACTICE

Activities like the following would be suitable:

1. (a) $246 = 200 + 40 + 6$ (c) $748 = 700 + \_\_ + 8$
   (b) $587 = \_\_ + 80 + 7$ (d) $6 + 300 + 50 = \_\_$
2. (a) Write the number made up of 6 hundreds, 4 tens, 8 units.
   (b) Write the number made up of 3 units, 2 hundreds, 8 tens.
3. In the number 763, the 6 is worth _____ .
4. Make the largest/smallest number with these digits: 3 1 8.

EVALUATION

The following ideas are fundamental to a child's understanding of the concept of place value:

1. Our place-value system uses ten digits — 0, 1, 2, 3, 4, 5, 6, 7, 8, 9
2. A digit's place in a number determines its value, e.g. in 756 the 5 stands for 5 lots of 10 (5×10).
3. The importance of zeros as place holders, e.g. <u>500.</u>
4. Digits have a dual value. In 476, the 7 has the absolute value of 7 and, because of its position in the number, it is identified as 7 tens or 70. Therefore, it accounts for 70 of the total 476.

CONCEPT Place Value — renaming

CONCRETE MATERIALS

1. M.A.B. 10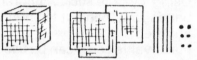
   1,356 = 1,356 units
   　　　　13 hundreds + 56 units
   　　　　135 tens + 6 units etc.
   How many tens are in 1,356?
   There are the obvious five, plus the tens in the hundreds, plus the tens in the thousand; altogether — 135 tens.
2. Place-value cards can be broken up in many ways, e.g.

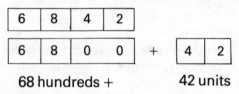

| 6 | 8 | 4 | 2 |

| 6 | 8 | 0 | 0 |　+　| 4 | 2 |

68 hundreds +　　　　　42 units

GAMES

'Place Value Concentration'.
The teacher works matching pairs of numbers on a set of blank playing cards. Cards are shuffled and turned face-down in the centre of the table. Children must find matching pairs. When found, they are displayed in front of the player.

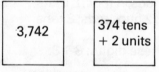

　　3,742　　　　374 tens + 2 units

Matching pair

FURTHER PRACTICE

Children should progress to being able to see the pattern when renaming numbers, e.g. if we want to know how many tens are in 4,768 we read from the left to the designated place — 4,768 so there are 476 tens in the number. How many thousandths are there in 46·257? 46,257 thousandths.

EVALUATION

Examples:
1. Section F
   26,347 = 263 _____ + 47 _____
   　　　 = 2634 _____ + 7 _____
   　　　 = 2 _____ + 634 _____
   　　　 + 7 _____

2. Section G
   63·49 = 63 _____ + 49 _____
   　　 = 9 _____ + 634 _____
   　　 = 6 _____ + 349 _____

3. Rename 3,817 five ways.

CONCEPT Place Value — further development

CONCRETE MATERIALS

When a child understands the concept of place value to hundreds, the same materials and ideas can be used to extend the study of place value to millions and thousandths.

1. **Place-value cards**

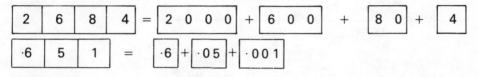

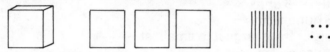

2. **M.A.B. 10**
   If the block = 1, 1·397 can be shown as:

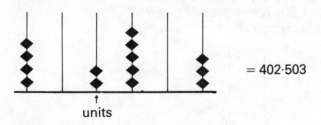

3. **Abacus**
   The three-spike interlocking abaci can be joined to extend the place-value range.

   When using an abacus, it is always necessary to indicate which place is the units. The units place is the key to the pattern of the place-value system (see Pattern and Order).

   Digits to the left of the units place are groupings of units, and places to the right show the fractional parts of a unit.

= 402·503

↑
units

GAMES

1. 'Decicus'
2. 'Wipe-Out'

EVALUATION

The following examples would be a reasonable expectation when evaluating children's understanding of place value in those sections:

Section F

1. Extend 206,564
2. $200 + 7 + 600,000 + 40 + 3,000 + 90,000 =$
3. $876 = $ _____ hundreds + _____ units
4. $186.4 = $ _____ tens + _____ tenths

5. $0.3 = \dfrac{\Box}{\Box}$ $\dfrac{9}{10} = \cdot \Box$ $6\dfrac{4}{10} = \Box \; \Box$

Section G

1. $90,000 = 9 \times$ _____
   $= 900 \times$ _____
   $= 90,000 \times$ _____
   $= 90 \times$ _____

2. $36,482 = 36$ _____ $+ 48$ _____ $+ 2$ _____
   $= 3,648$ _____ $+ 2$ _____
   $= 82$ _____ $+ 364$ _____
   $= 36,482$ _____

3. Change to decimals $\dfrac{7}{10}$ $\dfrac{8}{100}$ $14\dfrac{19}{100}$
4. 9 tenths + 7 hundredths =
5. 23 hundredths + 6 units =
6. 15 tens + 15 hundredths =
7. Arrange in order from largest to smallest —
   $\cdot 6$ $\dfrac{7}{100}$ 800   80·0   ·06   8·8

Section H

1. Extend 7,648,917
2. $16.843 = 16$ _____ $+ 843$ _____
   $= 84$ _____ $+ 16$ _____ $+ 3$ _____
   $= 168$ _____ $+ 43$ _____
   $= 16,843$ _____
3. Change to decimals —
   $\dfrac{7}{10}, \quad \dfrac{7}{100}, \quad \dfrac{7}{1000}, \quad 6\dfrac{84}{1000}, \quad 19\dfrac{1}{1000}$

4. Change to fractions — 0·87, 0·586, 4·23, 18·001
5. Make this number 1,000 times greater — 46·83
6. (a) What number is represented on the abacus?

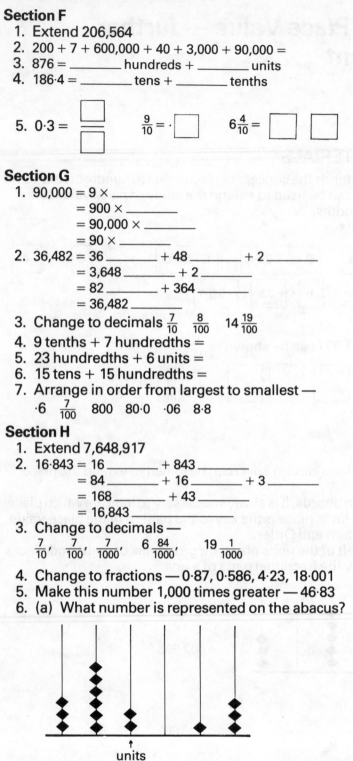

units

(b) What number is $\dfrac{1}{100}$ more than the number shown?

CONCEPT Bases Other than Base 10

CONCRETE MATERIALS

1. M.A.B. 3, 4 and 5. Let children see that other bases have minis, longs, flats and blocks as in base 10, but the system of grouping determines the respective values, e.g. base 4

mini	long	flat	

 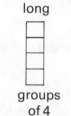 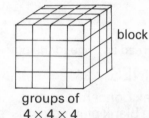

units	groups of 4	groups of 4 × 4	groups of 4 × 4 × 4

2. Chip Trading. Any coloured discs can be used and any value placed on them, e.g. if you get four whites, change for a blue or if you get four blues, change for a red etc.
    Emphasizes grouping in other bases.

GAMES

1. 'Make a Flat'. Play in different bases, e.g. in base 5, the first to roll 25 wins.
2. 'Make a Block' would be more appropriate in base 3 and base 4.

FURTHER PRACTICE

Let children discuss and invent new digits for bases above ten. This enables children to view the ten digits from a different perspective. Change from base 10 to other bases, e.g. $32 = 112_{five}$ and $8 = 1000_{two}$

and from other bases to base 10, e.g. $40_{five} = 20$ and $222_{three} = \boxed{}$

Count in other bases.

EVALUATION

A study of other bases should add greater depth and understanding to a child's knowledge of base 10 and the place-value system.
    Many children appear to understand the place-value system but this can be due to a familiarity with constant counting activities and usage of base 10. Children should see that the base of any place-value system is determined by the method of grouping in that system.

CONCEPT Index Notation

CONCRETE MATERIALS

Index notation is a shorthand way of writing large numbers.

Using index notation
$$1,000,000 = 10 \times 10 \times 10 \times 10 \times 10 \times 10$$
$$= 10^6 \leftarrow \text{index or exponent}$$
$$\uparrow \text{ base}$$

5^4 is read as: 'five to the power four', or 'five to the fourth power'.

GAMES

'Index Concentration'
    On blank playing cards, prepare sets of three cards as follows:

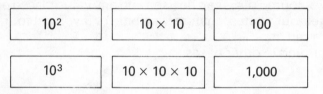

| 10^2 | 10×10 | 100 |
| 10^3 | $10 \times 10 \times 10$ | 1,000 |

Do the same for 10^4, 10^5, 10^6, 10^1.
    The cards are spread face down in the centre. Each child picks up three cards, trying to get three that match. Any matching cards are turned face up in front of the player.

FURTHER PRACTICE

Base 10 work could be linked to serial multiplication and division, e.g.

$100 \times 100 = 10,000$	$1,000 \div 100 = 10$
$10^2 \times 10^2 = 10^4$	$10^3 \div 10^2 = 10^1$
$10 \times 100 = 1,000$	$1,000,000 \div 1,000 = 1,000$
$10^1 \times 10^2 = 10^3$	$10^6 \div 10^3 = 10^3$

Children should be able to make the generalization that when we multiply indices with the same base we add the indices, and when we divide indices with the same base, we subtract.

EVALUATION

Typical examples would be:

$10 \times 10 \times 10 = 10^{\square}$ $10^4 \times 10^3 = 10^{\square}$

$4^5 = 4 \times 4 \times 4 \times 4 \times \square$ $5^3 \div 5^2 = \square$ or $\square$

$5^3 = 5 \times 5 \times 5$ or $\square$ $10,000 = 10^{\square}$

PROCESSES

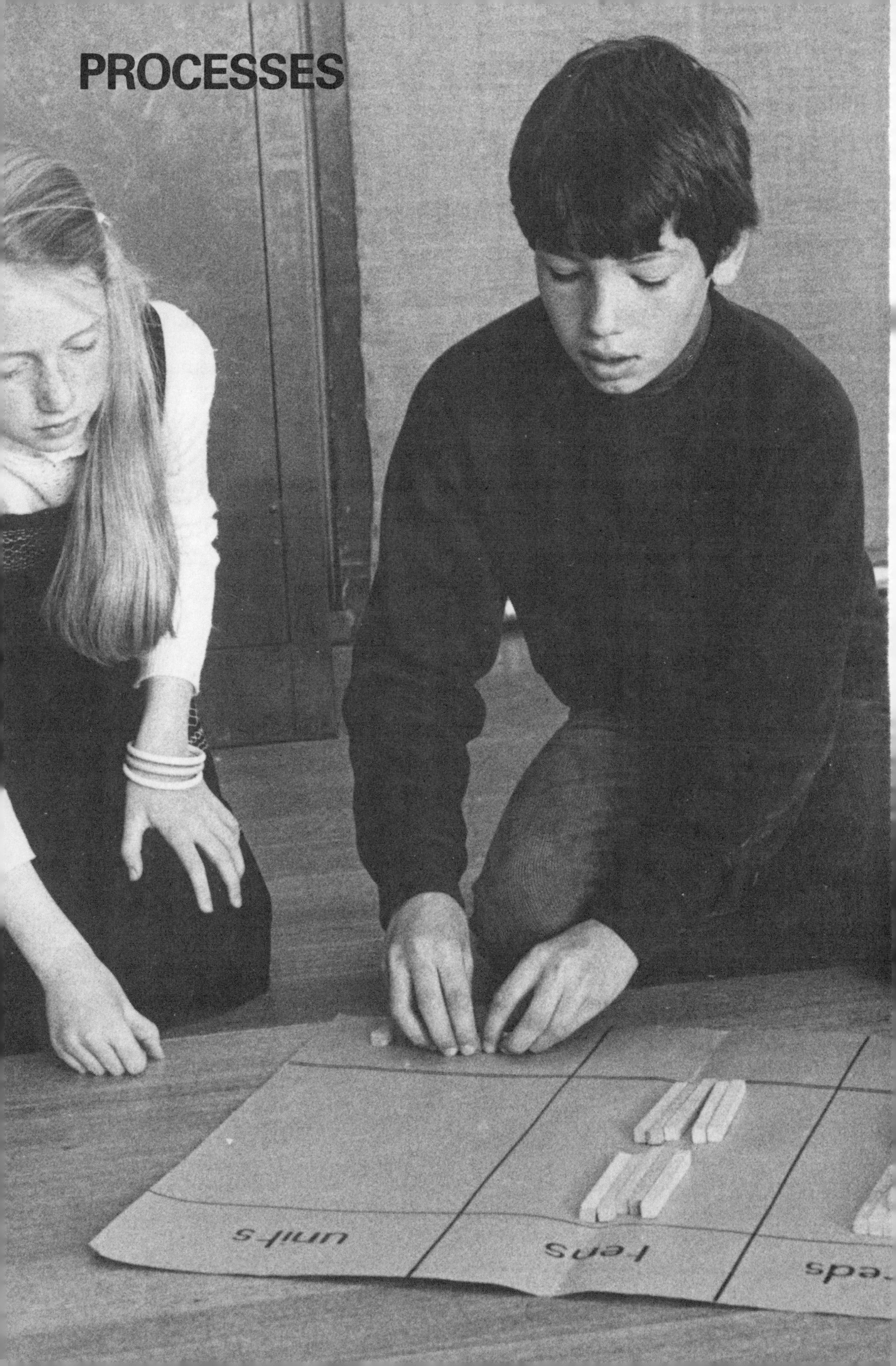

Developing the Four Processes — Algorithms

It has been common practice in Victorian schools to introduce the algorithms through stages of refinement using extended horizontal setting out. This has been an unnecessarily drawn-out process resulting in confusion for many children.

The approach outlined in this book develops the final refinement from the outset through the use of M.A.B.10, icy-pole sticks and place-value charts.

STEPS

1. Have children represent the numbers on the place-value charts with M.A.B. 10 or icy-pole sticks, e.g.
   Addition: 136 + 245

	hundreds	tens	units
(136)	□	|||	:·:
(245)	□ □	||||	::·

   N.B. Addition, multiplication and subtraction are done on place-value charts.
   Short division (sharing) is not suited to this setting out.

2. As children join the two numbers together, the teacher records the setting out on the chalkboard or a large sheet of paper.
3. The formal recording is linked to the physical manipulations done by the children, with discussion of each step being essential.
4. In later sessions the children can record their own workings with the materials.
5. Children use the materials until they feel confident to complete the algorithm without materials. Of course, the amount of time taken to become confident without materials varies from child to child.

CONCEPT Addition Algorithm

CONCRETE MATERIALS

Icy pole sticks and M.A.B. 10 with place-value charts

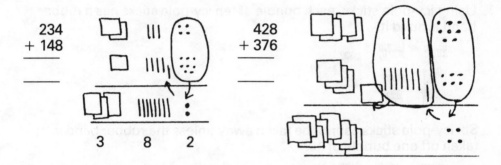

The teacher could record the setting out on paper or the chalkboard after children complete each example.

Discuss each problem with the children, asking them questions such as, 'What was the one we put with the tens?' Children can record the workings themselves after the first session.

GAMES

1. 'Make a Flat' (M.A.B. 10). Play as a prerequisite to learning this concept.
2. 'Make a Bundle'. As above, but use icy-pole sticks instead.
3. 'Build the Greatest Sum'

FURTHER PRACTICE

Stages of development (Section E):

1. 324 + 145 (no regrouping)
2. 246 + 238 (regrouping of units)
3. 253 + 364 (regrouping of tens)
4. 457 + 268 (regrouping of tens and units)

Children will use materials while it assists their understanding. The teacher will see the 'signs' when children no longer need materials.

EVALUATION

Make sure that children verbalize each physical manipulation while performing the algorithm with M.A.B.10 or icy-pole sticks. This gives the teacher an insight into the child's thinking and understanding.

CONCEPT Subtraction Algorithm — Decomposition

CONCRETE MATERIALS

N.B. Only the number you are 'taking away from' is represented with materials, e.g. 84 − 36.

1. Using icy-pole sticks, each bundle of ten icy-pole sticks has a rubber band around it.

Six icy-pole sticks cannot be taken away unless the rubber band is taken off one bundle of 10.

Six can now be taken away and then three bundles of 10 removed.
   The teacher records children's working in the first session on the chalkboard or large sheets of paper.

2. M.A.B. 10
   If six units cannot be removed, one 10 is exchanged for ten units.
   Six units then three tens are removed.

$$
\begin{array}{r} 84 \\ -36 \\ \hline \\ \hline \end{array}
\qquad
\begin{array}{r} 7\,\overset{1}{8}\,4 \\ -3\,6 \\ \hline \\ \hline \end{array}
$$

Icy-pole sticks are better than M.A.B. 10 initially because it is easier to remove a rubber-band around a bundle of 10 than to exchange one 10 for ten units. This requires working with material that wasn't represented in the original number because ten more units must be taken from the box of M.A.B.

GAMES

1. 'Build the Greatest/Smallest Difference'
2. 'Break a Flat/Bundle'

FURTHER PRACTICE

Stages of development:

1. 86 − 34 (no renaming)
2. 486 − 235 (no renaming)
3. 84 − 36 (one renaming step)
4. 327 − 149 (two renaming steps)
5. 400 − 236
    402 − 137 (examples involving zero)

EVALUATION

Children should realize that the different aspects of subtraction can all be solved using this decomposition algorithm:

1. 436 less 294 (take-away)
2. What is the difference between 206 and 97? (difference aspect)
3. What must I add to 346 to get 501? (complementary addition)

CONCEPT Short Division Algorithm — partition (equal shares method)

CONCRETE MATERIALS

Icy-pole sticks and M.A.B. 10
369 ÷ 3 means 369 divided into three equal shares.

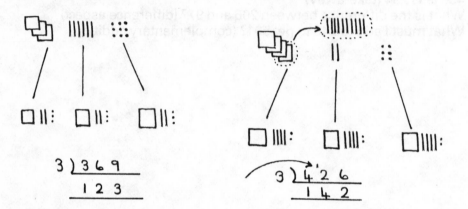

$$3 \overline{) 3\ 6\ 9}$$
$$1\ 2\ 3$$

$$3 \overline{) 4\ \overset{1}{2}\ 6}$$
$$1\ 4\ 2$$

100 must be changed into (renamed as) 10s to enable equal shares; then twelve 10s are shared between 3.
The carrying figures are simply numbers being renamed.

N.B. Whenever the hundreds or tens can't be shared equally, the part that can't be shared is renamed.

FURTHER PRACTICE

Development:
1. equal shares without renaming, e.g. 369 ÷ 3
2. renaming of hundreds only, e.g. 426 ÷ 3
3. renaming of tens only, e.g. 560 ÷ 5
4. renaming of hundreds and tens, e.g. 435 ÷ 3

Once children understand the concept of 'carrying' or renaming, the material will not be necessary and the degree of difficulty can be increased (e.g. to include thousands).

EVALUATION

Children should understand that '426 shared equally between 3', and 'how many 3s are there in 426?' gives the same numerical answer. Therefore the short division algorithm as developed above is appropriate to solve both partition (equal shares) and quotition (how many) problems.

Division of decimals (e.g. 684.5 ÷ 5 — under Fractions in Section H) is the same concept. However when the units cannot be shared equally, they are renamed as tenths, or tenths renamed as hundredths etc.

CONCEPT Short Multiplication Algorithm

CONCRETE MATERIALS

Use either icy-pole sticks or M.A.B. 10.
3 × 36 means 3 lots of 36.

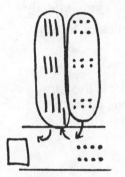

$$\begin{array}{r} 36 \\ \times 3 \\ \hline 108 \\ \hline \end{array}$$

1. Three lots of six equal 18, which is one ten and eight units left over.
2. Three lots of 30 equals 90, plus the extra 10 we made, making 100 altogether.

GAMES

'Roll a Product'

FURTHER PRACTICE

Stages of development:
Section E   86 × 4   ,   5 × 42
Section F   132 × 8   ,   11 × 264
Section G   3,618 × 7   ,   12 × 5,706

Make sure many examples are given in problem-solving form, e.g.
I had four boxes with 75 oranges in each. How many oranges do I have altogether?

EVALUATION

Allow children to use concrete materials as long as they require them, therefore you might evaluate this concept at two levels:
1. A child can complete short multiplication problems with materials.
2. A child does not require materials (multiplication tables are good enough to enable quick accurate calculations).

CONCEPT Long Multiplication Algorithm

CONCRETE MATERIALS

This is one concept in which concrete materials do not assist a child's understanding, in fact for many children they cause confusion.

Once a child has learnt to do short multiplication using M.A.B. 10 and icy-pole sticks, the following stages of development are very effective.

1. Give examples where children multiply by multiples of 10, e.g.

$$\begin{array}{ccc} 26 & 34 & 72 \\ \times\,10 & \times\,20 & \times\,30\ \text{etc.} \\ \hline \end{array}$$

Lead children to see that when we multiply by 30 it is the same as multiplying by 10 then by 3. Multiplying by 10 adds a zero to the number, so we add a zero then multiply by 3, e.g.

$$\begin{array}{r} 72 \\ \times\,30 \\ \hline 216\,0 \\ \hline \end{array}$$

multiply   multiply by 10
by 3

2.
$$\begin{array}{r} 74 \\ \times\,28 \\ \hline 592 \\ 1480 \\ \hline 2072 \\ \hline \end{array}$$

should be seen as three operations:

(a) 74 × 8 (short multiplication algorithm)
(b) 74 × 20 (previously developed)
(c) total parts (a) and (b)

3. Make sure children make up 'real life' situations for long multiplication problems and vice versa.

BASIC PROPERTIES

Developing the Basic Laws of Mathematics

The following laws should not be 'formally' taught, but when situations naturally arise in discussion, this is the ideal time to ensure the following concepts are thoroughly understood.

1. Commutative law of addition — $4 + 3 = 3 + 4$ (Order doesn't
2. Commutative law of multiplication — $3 \times 5 = 5 \times 3$ matter.)
3. Associative law of addition —

$$(2 + 3) + 4 = 2 + (3 + 4)$$
$$5 + 4 = 2 + 7$$
$$9 = 9$$

4. Associative law of multiplication —

$$(2 \times 3) \times 4 = 2 \times (3 \times 4)$$
$$6 \times 4 = 2 \times 12$$
$$24 = 24$$

N.B. Because we have a binary system we can only work with two numbers at one time. Thus the need for the associative or grouping law.

5. The Distributive Law of:
   (a) multiplication over addition — 6×8

$$= 6 \times (5 + 3)$$
$$= (6 \times 5) + (6 \times 3)$$
$$= 30 + 18$$
$$= 48$$

   (b) multiplication over subtraction — $4 \times 8 = 4 \times (10 - 2)$

$$= (4 \times 10) - (4 \times 2)$$
$$= 40 - 8$$
$$= 32$$

   (c) division over addition — $36 \div 6$

$$= (30 + 6) \div 6$$
$$= (30 \div 6) + (6 \div 6)$$
$$= 5 + 1$$
$$= 6$$

6. Law of addition of zero $8 + 0 = 8$
7. Law of multiplication of zero $3 \times 0 = 0$
8. Law of multiplication by one $3 \times 1 = 3$
9. Equals added to equals (the sums are equal), e.g.:
   (a) $14 + 6 = 20$
   (b) $14 + 6 + 5 = 20 + 5$ (add 5 to both sides)

10. Equals added to unequals (the original difference remains the same), e.g.:
    (a) $95 - 26 = 69$
    (b) $100 - 31 = 69$ (add 5 to both sides)
    This idea can often make a problem easier to do mentally, e.g. $996 - 269$.
    Add 4 to both sides, $1,000 - 273 = 727$
    Children must understand this idea if they use the equal additions algorithm for subtraction.

11. Equals subtracted from unequals (the original difference remains the same), e.g.:
    (a) $8 + 6 = 2$
    (b) $6 - 4 = 2$
        (2 is subtracted from both the 6 and the 4)
    (c) $105 - 68$
    (d) $100 - 63 = 37$ (subtract 5 from both sides)

Here is an example of how a mathematical law might be developed.

SECTION B — COMMUTATIVE LAW OF MULTIPLICATION $(a \times b = b \times a)$

When learning the basic law of multiplication, a child might discover that three lots of four is the same as four lots of three.
   Activities that lead to this discovery could include:
1. Mathematical balance, e.g.

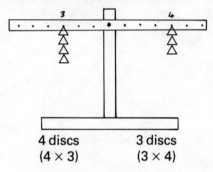

4 discs   3 discs
(4×3) (3×4)

2. Arrays — this array can be seen as three lots of four or four lots of three.

3. Activity sheets, e.g.
   $4 \times 2 =$ $2 \times 4 =$
   $3 \times 2 =$ $2 \times 3 =$
   Children generally tell you why many of the answers are the same.
      By Section E, a child will see that the law also applies to larger numbers, e.g.
         $3 \times 126 = 126 \times 3$
   3 lots of 126 $=$ 126 lots of 3.
   The commutative law of multiplication is extremely useful when memorizing multiplication tables.
      If a child has learnt 12×2 (2 times table), they have also learnt 2×12 (12 times table).
      By Section G, children will see that $3 \times \frac{1}{4}$ (three lots of $\frac{1}{4}$) is a simpler concept than $\frac{1}{4} \times 3$ ($\frac{1}{4}$ of 3) and $\cdot 4 \times 3$ ($\frac{4}{10}$ of 3) is more easily understood as $3 \times \cdot 4$ (three lots of $\cdot 4$).

By section H, children can be given problems like the following:

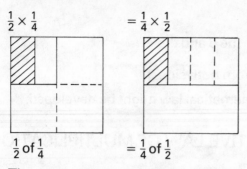

$$\frac{1}{2} \times \frac{1}{4} \qquad = \frac{1}{4} \times \frac{1}{2}$$

$$\frac{1}{2} \text{ of } \frac{1}{4} \qquad = \frac{1}{4} \text{ of } \frac{1}{2}$$

The term commutative need not be used at any stage in this development. Children need only know that the **order** in multiplication does not matter.

Children should know that addition is also commutative $(3 + 6 = 6 + 3)$, but that subtraction and division are not $(6 - 3 \neq 3 - 6, 12 \div 3 \neq 3 \div 12)$.

FRACTIONS

CONCEPT Fraction as a Relationship between Two Whole Numbers, e.g. $\frac{1}{2}$, $\frac{1}{3}$

CONCRETE MATERIALS

1. Paper folding with coloured squares, e.g. fold into quarters, reverse the white side and paste to highlight $\frac{1}{4}$ or one out of four equal parts.
2. Use an S.A. 64 or Wainwright fraction kit to highlight the concept of fractions. Build a 'Big Mac' with different 'layers' of fractions, e.g. $\frac{1}{2}$ s, $\frac{1}{3}$ s, $\frac{1}{4}$ s, $\frac{1}{5}$ s etc.
3. Cuisenaire rods show the relationship between two numbers, e.g. $\frac{1}{3}$, $\frac{2}{5}$

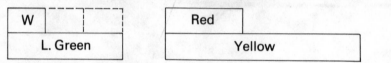

4. Use cut-out shapes and ask the children what fraction is shaded. Vary the shape and fraction.

5. Geoboards. Show $\frac{1}{4}$, $\frac{1}{3}$, $\frac{1}{2}$ etc.

GAMES

1. 'Fraction dominoes' (commercially available)

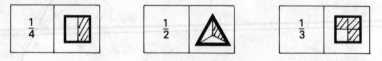

2. 'The Hamburgler Game'

FURTHER PRACTICE

1. Opportunities may arise in the classroom to reinforce the concept, e.g. windows divided into four equal parts, but *make sure* the objects are divided into *equal* parts.
2. Children should be able to identify shaded parts, e.g.

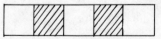

3. Ask the children to:
   (a) complete the number lines, e.g.

   (b) circle the largest fraction —
   $\frac{1}{4}, \frac{1}{2}, \frac{1}{10}$ and $\frac{1}{9}, \frac{3}{9}, \frac{8}{9}$ etc.

EVALUATION

In addition to children being able to name fractional parts, it is important that they continue to explain what is actually meant by a fraction, e.g. $\frac{1}{5}$ means one out of five equal parts.

CONCEPT Fraction as an Operator, e.g. $\frac{1}{4}$ of 12

CONCRETE MATERIALS

1. Discrete materials, e.g. counters, discs.
   A child is asked to work out $\frac{1}{3}$ of 6. He/she puts out six objects:
   and divides the group into thirds.

   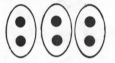

   $\frac{1}{3}$ of 6 means one of the three equals groups, i.e. 2.

2. Cuisenaire rods.
   If white equals 1, brown equals 8 and red equals 2.

	Brown		
Red			

   $\frac{1}{4}$ of 8 = (one of the four equal rods or 2)

   $\frac{1}{4}$ of 8 = 2

   $\frac{2}{4}$ of 8 = 4 (or two of the four equal parts)

   $\frac{3}{4}$ of 8 = 6

   $\frac{4}{4}$ of 8 = 8

GAMES

1. 'Operation Bingo'
   All children have base cards. One child is the caller and turns over the cards, e.g. $\frac{1}{2}$ of 10. Children with the number 5 cover it with a counter.
   The first child to cover the entire card wins.

3	1	8
4	2	5

FURTHER PRACTICE

Use discrete materials in everyday situations, e.g. $\frac{1}{5}$ of five lollies, $\frac{1}{2}$ of twenty children etc.
Allow children to use materials as long as necessary.

$\frac{1}{3}$ of 9 =

$\frac{2}{3}$ of 9 =

$\frac{3}{3}$ of 9 =

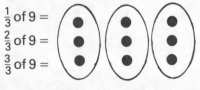

If a child has found $\frac{1}{3}$ of a number, let them continue the pattern by finding $\frac{2}{3}$ and then $\frac{3}{3}$ of the number.

Link this concept to automatic response activities.

If a child knows $3 \times 5 = 15$, encourage her/him to use $\frac{1}{3}$ of $15 = 5$ and $\frac{1}{5}$ of $15 = 3$.

EVALUATION

Make sure that children are not forced to work without materials too soon. As with most concepts, children should be able to use materials when being evaluated. It is a child's understanding that is being evaluated, not his/her knowledge of number facts.

CONCEPT Equivalent Fractions

CONCRETE MATERIALS

1. Children shade in half of their piece of paper, and each successive fold in half reveals other equivalent fractions for $\frac{1}{2}$. This method can also be used to show equivalent fractions for $\frac{1}{3}$ and $\frac{1}{5}$ 'families', e.g.

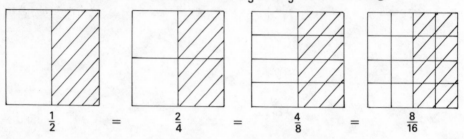

$$\frac{1}{2} \quad = \quad \frac{2}{4} \quad = \quad \frac{4}{8} \quad = \quad \frac{8}{16}$$

2. Place the fraction wall under the plastic of the acetate board (see page 84) and ask the children to fill it in, e.g. find other names for $\frac{1}{2}$

halves	$\frac{-}{2}$
thirds	$\frac{-}{3}$
quarters	$\frac{-}{4}$
fifths	$\frac{-}{5}$
sixths	$\frac{-}{6}$

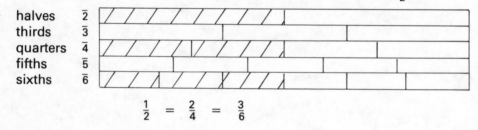

$$\frac{1}{2} \quad = \quad \frac{2}{4} \quad = \quad \frac{3}{6}$$

GAMES

1. 'Fracto' (equivalent fraction bingo)
2. 'Equivalent Fraction Dominoes'
3. 'Equivalent Fraction Snap'

FURTHER PRACTICE

1. Conduct an equivalent fraction hunt. Link it with counting/pattern and order (see Counting, Pattern and Order, p.22).
2. Lead children to make generalizations about their understanding of equivalence, e.g. equivalence is maintained if we multiply or divide the numerator and denominator by the same number, e.g.

$$\frac{1 \times 3}{4 \times 3} = \frac{3}{12},$$

$$\frac{6 \div 6}{18 \div 6} = \frac{1}{3}$$

Include in pattern work sessions.

3. Show children how to reduce to the lowest term or simplest name, e.g.

$$\frac{3 \div 3}{15 \div 3} = \frac{1}{5}$$

EVALUATION

Many children who understand the concept of equivalent fractions will not perform particularly well in this area until their automatic response is satisfactory.

FRACTION WALL

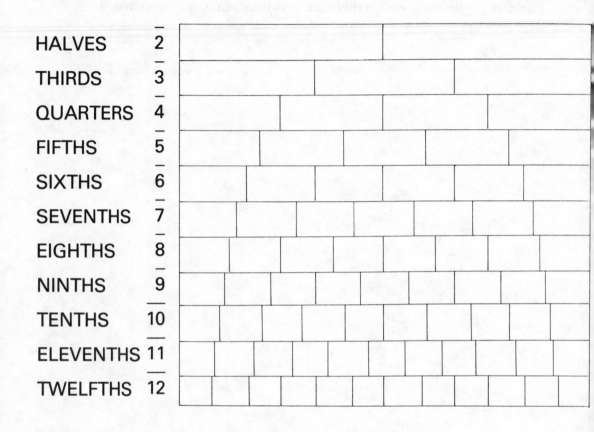

HALVES	$\frac{}{2}$
THIRDS	$\frac{}{3}$
QUARTERS	$\frac{}{4}$
FIFTHS	$\frac{}{5}$
SIXTHS	$\frac{}{6}$
SEVENTHS	$\frac{}{7}$
EIGHTHS	$\frac{}{8}$
NINTHS	$\frac{}{9}$
TENTHS	$\frac{}{10}$
ELEVENTHS	$\frac{}{11}$
TWELFTHS	$\frac{}{12}$

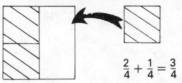

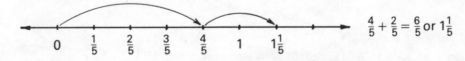

CONCEPT Addition of Vulgar Fractions — like denominators

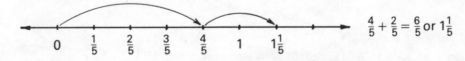

N.B. Subtraction of vulgar fractions can be developed with the same materials and activities. Because subtraction is the inverse of addition, if a child can put $\frac{3}{5} + \frac{1}{5}$ together to make $\frac{4}{5}$, then $\frac{1}{5}$ can be taken away again to show $\frac{4}{5} - \frac{1}{5} = \frac{3}{5}$.

CONCRETE MATERIALS

1. Using fraction kits (S.A. 64 and Wainwright), place fractional parts on a unit. The children should read and record the equations, e.g.

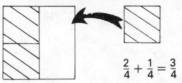

$$\frac{2}{4} + \frac{1}{4} = \frac{3}{4}$$

2. Make a 'fraction cake' and let the children explore the different layers of the cake, recording all possible equations.

$\frac{7}{8} + \frac{1}{8} = \frac{8}{8}$ or 1

$\frac{3}{8} + \frac{4}{8} = \frac{7}{8}$

$\frac{1}{8} + \frac{4}{8} = \frac{5}{8}$ etc.

3. Cuisenaire rods. If orange = 1, then $\frac{4}{10} + \frac{6}{10} = 1$

Orange	
Pink	Dark Green

4. Number lines

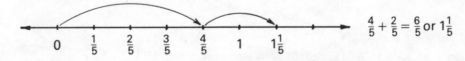

$\frac{4}{5} + \frac{2}{5} = \frac{6}{5}$ or $1\frac{1}{5}$

GAMES
'Build a Unit'

FURTHER PRACTICE
Confine examples to the range of the S.A. 64 and Wainwright fraction kits. Never resort to rules for this or any concept. This idea is a simple one for children to understand if they use their knowledge of the basic operation of addition while manipulating concrete materials.

EVALUATION
Make sure children continue to verbalize in simple language what happens when fractions are added or subtracted, e.g.

1. $\frac{2}{6} + \frac{3}{6} = \frac{5}{6}$

   $\frac{2}{6}$ put with $\frac{3}{6}$ equals $\frac{5}{6}$

2. $\frac{7}{8} - \frac{3}{8} = \frac{4}{8}$

   $\frac{7}{8}$ take away $\frac{3}{8}$ leaves $\frac{4}{8}$

CONCEPT Addition of Vulgar Fractions — unlike denominators

The same approach is used to develop subtraction of vulgar fractions.

CONCRETE MATERIALS

Use S.A. 64 and Wainwright fraction kits.
Unlike fractional parts, these do not show an immediate answer when joined together.

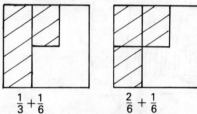

$$\frac{1}{3}+\frac{1}{6} \qquad \frac{2}{6}+\frac{1}{6}$$

Let children see that if $\frac{1}{3}$ is renamed as $\frac{2}{6}$, the answer is obvious.

Suggested setting out for Section G.

$$\frac{1}{3}+\frac{1}{6}$$
$$=\frac{2}{6}+\frac{1}{6}$$
$$=\frac{3}{6}\text{ or }\frac{1}{2}$$

Children must choose which fraction is to be renamed (equivalent fractions).

Even with a fraction kit this is a trial and error approach for some children in the initial stages.

Examples where both fractions must be renamed (Section H).

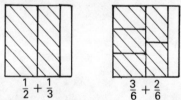

$$\frac{1}{2}+\frac{1}{3} \qquad \frac{3}{6}+\frac{2}{6}$$

Children try to rename one fraction at first but soon realize that they must find fractional parts that can cover *both* fractions, e.g. $\frac{1}{2}+\frac{1}{3}$.

$\frac{1}{2}$ is covered by $\frac{3}{6}$ and $\frac{1}{3}$ by $\frac{2}{6}$, so:

$$\frac{1}{2}+\frac{1}{3}$$
$$=\frac{3}{6}+\frac{2}{6}$$
$$=\frac{5}{6}$$

N.B. Do not rush children to 'see' how the lowest common denominators can be used. Children will see this concept in their own time and will eventually apply it with real understanding.

CONCEPT Addition and Subtraction of Decimals

CONCRETE MATERIALS

Use M.A.B. 10 and place-value charts.
Example using hundredths —

$$46\cdot18$$
$$+\ 28\cdot74$$

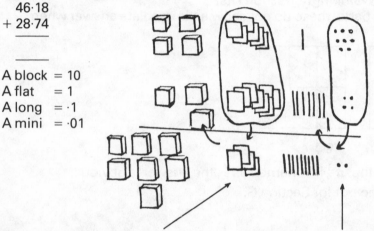

A block = 10
A flat = 1
A long = ·1
A mini = ·01

Fourteen units make a new ten with four units left over.

Twelve hundredths make a new tenth with two hundredths over.

Subtraction is treated using the decomposition method but the M.A.B. 10 is given appropriate values as with addition.

When thousandths are added or subtracted, the block equals one and examples with M.A.B. 10 are restricted to examples from units to thousandths.

GAMES

'Build the Greatest Sum' (addition)
'Build the Greatest Difference' (subtraction)
'Decicus'

FURTHER PRACTICE

Children generally do not need to use M.A.B. 10 for long before they realize that the ideas they learned when adding whole numbers still apply to the addition of decimals.

EVALUATION

In addition to examples like the following:

68·4 and	23·648
− 26·87	17·964
	+ 18·72

examples using hundredths should make a link with money, e.g.

$64·86
− 28·69
_____

_____

and thousandths should be linked to the metric system, e.g.

kg	km	m
2·468	3·742	2·758
+ 4·739	− 2·196	+ 3·647
_____	_____	_____

CONCEPT Multiplication of Decimal Fractions (e.g. In Section G — 3 × ·4, Section H — ·4 × ·5)

CONCRETE MATERIALS

1. Children need to realize the relationship between multiplication and addition, e.g.
   ·7 + ·7 + ·7 = 2·1, 3 × ·7 = 2·1
   Use number lines to demonstrate this, e.g. 3 × ·4 = 1·2

2. Cuisenaire rods can also be used, e.g.
   if orange = 1, 3 × ·4 = 1·2

orange (1)		red (·2)
pink (·4)	pink (·4)	pink (·4)

3. Use knowledge of vulgar fractions, e.g. $5 \times \frac{3}{10} = \frac{15}{10}$ or $1\frac{1}{2}$, 5 × ·3 = 1·5 or $1\frac{1}{2}$.

GAMES

'Spin a decimal'. The units are written on playing cards and decimals on a spinner. Each child takes a card, and then spins the spinner, e.g. a child picks the card 6 and then spins ·8. She/he has to multiply 6 × ·8 (4·8). The largest answer scores a point.

FURTHER PRACTICE

Lead children to discover that units times tenths will always give an answer showing tenths, e.g. 5 × ·7 = 3·5, 8·6 × 4 = 34·4, 642·7 × 3 = 1928·1, 4·8 × 26 = 124·8.

Check by estimating answers. Show the children how to round off to whole numbers, e.g. 35 × 6·4 can be rounded off to 6 × 35 = 210, so 35 × 6·4 = 224·0 is a reasonable answer.

EVALUATION

It is important for children to understand that:
1. units × tenths = tenths, e.g. 4 × ·6 = 2·4
2. tenths × tenths = hundredths, e.g. ·6 × ·7 = ·42
3. tenths × hundredths = thousandths, e.g. 3·4 × 6·48 = 22·032

Children should *always* estimate their answers to prevent silly slips in the placement of the decimal point.

Some children may prefer to count the places to the right of the decimal point, then place the decimal point in their answer.

CONCEPT Multiplication of Vulgar Fractions, e.g. $4 \times \frac{1}{5}$

CONCRETE MATERIALS

1. Children must understand that multiplication is really repeated addition, e.g. $3 \times \frac{1}{4}$ is the same as $\frac{1}{4} + \frac{1}{4} + \frac{1}{4}$.

    Use an S.A. 64 fraction kit to let children discover that

    3 lots of $\frac{1}{4} = \frac{3}{4}$

    $3 \times \frac{1}{4} = \frac{3}{4}$

    Children quickly realize, when using materials, that because they can add 'like' fractions, multiplication is not really a new concept.

2. Number lines can also be used to demonstrate this, e.g. $4 \times \frac{1}{5} = \frac{4}{5}$

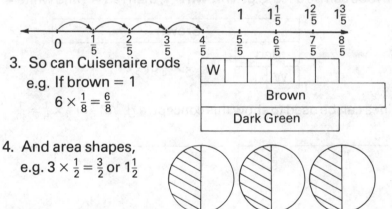

3. So can Cuisenaire rods

    e.g. If brown = 1

    $6 \times \frac{1}{8} = \frac{6}{8}$

4. And area shapes,

    e.g. $3 \times \frac{1}{2} = \frac{3}{2}$ or $1\frac{1}{2}$

GAMES

'Vulcation'

FURTHER PRACTICE

Make sure children use their knowledge of the commutative law of multiplication to give greater flexibility to solving equations, e.g. $\frac{1}{4} \times 9$ is more easily read as 9 lots of $\frac{1}{4}$ or $9 \times \frac{1}{4} = \frac{9}{4}$ or $2\frac{1}{4}$ and $27 \times \frac{1}{3}$ (27 lots of $\frac{1}{3}$) can be written as $\frac{1}{3} \times 27$ ($\frac{1}{3}$ of 27).

EVALUATION

Do not give children examples that require lengthy calculations. This just becomes 'busy' work. The following would be suitable examples:

$3 \times \frac{1}{4}, \frac{1}{5} \times 9, 4 \times \frac{2}{5}, \frac{5}{6} \times 10$.

Confine examples to known multiplication tables.

CONCEPT Multiplication of Vulgar Fractions, e.g. $\frac{1}{3} \times \frac{1}{4}$

CONCRETE MATERIALS

1. Use an S.A. 64 fraction kit (or mixed Wainwright), e.g. $\frac{1}{2}$ of $\frac{1}{3} = \frac{1}{6}$

A child first shows a third, then places a block on top which is half of one third.

2. Area diagrams can also be used. These are the same as the fraction kit above, but not three-dimensional.

3. Cuisenaire rods can be used, e.g. if brown = 1, then red = $\frac{1}{4}$ and white = $\frac{1}{2}$ of $\frac{1}{4}$ or $\frac{1}{8}$.

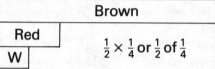

$\frac{1}{2} \times \frac{1}{4}$ or $\frac{1}{2}$ of $\frac{1}{4}$

4. Number lines can be used to show this concept, e.g. $\frac{1}{2}$ of $\frac{1}{5}$ or $\frac{1}{2} \times \frac{1}{5} = \frac{1}{10}$

FURTHER PRACTICE

Children should realize that to verbalize $\frac{1}{2}$ of $\frac{1}{4}$ as 'half lots of a quarter' is not easily understood. Therefore we shorten it to 'half of a quarter', a statement that is more closely aligned to natural language.

Children should be able to make up 'real-life stories' for equations such as 'I took a third of the pie and shared it with my friend, so we had half of the third each or one sixth of the pie each.' ($\frac{1}{2} \times \frac{1}{3}$)

EVALUATION

After children have used materials to learn this concept, they quickly see a pattern that makes the further use of materials unnecessary.

In all their examples the top numbers (numerators) are multiplied and the bottom numbers (denominators) are mutliplied, e.g.

$\frac{2}{3} \times \frac{1}{2} = \frac{2 \times 1}{3 \times 2} = \frac{2}{6}$

Let children discover this for themselves.

CONCEPT Division of Vulgar Fractions

CONCRETE MATERIALS

Use S.A. 64 or Wainwright fraction kits to demonstrate this concept.

Type 1 $2 \div \frac{1}{3}$

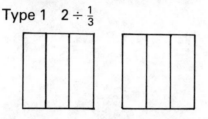

In 2, how many $\frac{1}{3}$s are there? (quotition division)

There are six thirds in 2, i.e. $2 \div \frac{1}{3} = 6$.

Type 2 $\frac{1}{2} \div 2$

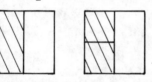

What is $\frac{1}{2}$ divided into two equal parts? (partition division)

$\frac{1}{2} \div 2 = \frac{1}{4}$

GAMES

'What's My Share?' (partition division). Write the fractions $\frac{1}{2}$, $\frac{1}{3}$ and $\frac{1}{4}$ four times each on playing cards, making twelve cards in all. Children in turn take a card and roll a six-sided die to determine how many parts the fraction is divided into, e.g.

$\frac{1}{2} \div 3 = \frac{1}{6}$

The largest fractional share wins a point.

FURTHER PRACTICE

N.B. Division of a fraction by a fraction, e.g. $\frac{1}{4} \div \frac{1}{3}$, is left until secondary school, however some children can use Type 1 problems to solve simple examples, e.g. $\frac{1}{3} \div \frac{1}{6}$ In $\frac{1}{3}$ how many sixths are there?

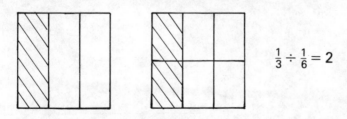

$\frac{1}{3} \div \frac{1}{6} = 2$

93

EVALUATION

Wherever children attempt division of fractions they should mentally try to verbalize the problem to determine which aspect of division (quotition or partition) gives the statement meaning, e.g. $\frac{1}{4} \div 2$.

1. A child tries quotition ('In a quarter how many twos are there?' — unsatisfactory).
2. He/she then tries partition ('One quarter shared between two people means they each receive $\frac{1}{8}$, so $\frac{1}{4} \div 2 = \frac{1}{8}$').

CONCEPT Percentage as a Fraction

CONCRETE MATERIALS

It is important that a child sees vulgar fractions, decimal fractions and percentages as three ways of expressing the same idea, e.g. $\frac{1}{4} = \cdot25 = 25\%$.

1. A counting frame (100 beads) can be used to illustrate this, e.g. 5 beads out of 100 $= \frac{5}{100}$ or 5%.

2. A 10 × 10 grid can also be used, e.g. shade in ten squares. What fraction have you shaded? $\frac{10}{100}$ or $\frac{1}{10} = 10\%$

   By definition the bead frame, or 10 × 10 grid, represents 100%. Per cent means 'per hundred' or 'parts of a hundred'. 25 per cent means twenty-five per hundred or twenty-five out of every hundred.

GAMES

1. 'Deci Maid'
2. 'Percentage Concentration'

FURTHER PRACTICE

1. Since each dollar is made up of 100 cents, money can be used to show the relationship between fractions, decimals and percent, e.g. one quarter of one dollar ($\frac{1}{4}$) is 25 cents or $0·25 or 25% of a dollar.
2. You can also compare measurements, e.g. length of 1 cm $= \frac{1}{100} = 1\%$ of metre.
3. Look for examples in everyday life when percentage is used, e.g. bank interest, exam results (secondary school), discounts, profits made, league team percentages etc.

EVALUATION

Examples like the following would be appropriate:

1. $\frac{1}{2} = $ ____ %, $\frac{1}{10} = $ ____ %, $\frac{7}{10} = $ ____%, $\frac{1}{8} = $ ____%, $\frac{1}{3} = $ ____%

2. $55\% = \frac{}{100} = \cdot$ ____

3. $1 - 0\cdot6 = $ ____ %

4. Express each percentage as a fraction in its lowest terms and then as a decimal fraction, e.g. $6\% = \frac{3}{50} = \cdot06$

   (a) 4%   (b) 1%   (c) 12%   (d) 25%

CONCEPT Percentage as an Operator

CONCRETE MATERIALS

Concrete materials should not be required to develop this concept as a
knowledge of previously learned concepts can be applied, viz:
1. fraction as an operator — $\frac{1}{5}$ of 10
2. concept of a percentage — 4 out of 100 is 4%
3. equivalent fractions — $\frac{5}{10} = \frac{1}{2}$
4. fractions can be written in many forms — $\frac{1}{4} = \cdot25 = \frac{25}{100} = 25\%$
5. multiplication of vulgar fractions — $\frac{1}{10} \times \frac{4}{1}$

GAMES

'Percentage Concentration'
Draw up matching pairs of cards, e.g.

| 2 | 5% of 40 |

Place the cards face down in the centre. Each child picks up two cards and
then puts down any matching pairs. The game ends when all the cards have
been used and the child with the most matching pairs wins.

FURTHER PRACTICE

Stages of development:
1. 25% of 40, $\frac{1}{4}$ of 40 = 10

2. 4% of $3, 4% is 4c in every dollar so 4c × 3 = 12c

3. 6% of 50, $\frac{6}{100} \times \frac{50}{1} = \frac{300}{100} = 3$

or $\frac{6}{100} \times \frac{50}{1} = 3$

We cancel because $\frac{6}{100} \times \frac{50}{1}$ can be written as $\frac{50}{100} \times \frac{6}{1}$

$\frac{1}{2} \times \frac{6}{1}$ (equivalent fractions).

EVALUATION

Sale catalogues would be useful to enable children to work out 'real life'
situations, e.g. '20% off all marked items'.

Reinforce this by looking for newspaper articles using percentages.
Photocopy these and discuss them in class.

Probability

Probability is the likelihood of an event occurring, e.g. if a coin is tossed, the probability of getting tails is one chance out of two ($\frac{1}{2}$).

A probability of 0 means the event can not occur, e.g. rolling an 8 on a six-sided die.

A probability of 1 means the event is sure to occur, e.g. when a two-headed coin is tossed, a head will appear.

Probability was originally used primarily to solve problems dealing with gambling. Today probability theory is used in all the sciences, and in medicine, economics and the manufacturing industry.

Probability of a particular event occurring can be worked out as follows:

$$\text{Probability (p)} = \frac{\text{Number of ways the event can occur}}{\text{Total number of possible events}}$$

The probability of throwing an even number on a six-sided die would therefore be p (even number) = $\frac{3}{6}$ or $\frac{1}{2}$.

ACTIVITIES

After teaching the children this concept, give them the following activities.

1. What is the probability of throwing a 6 on a normal die?
2. Use two dice and roll them 100 times. Add the figures shown on the two dice each time. Which number occurred most? Why?
3. List ten events with a probability of 0 and ten events with a probability of 1.
4. Toss two coins 100 times. What happens? What is the probability of tossing two heads? Why?

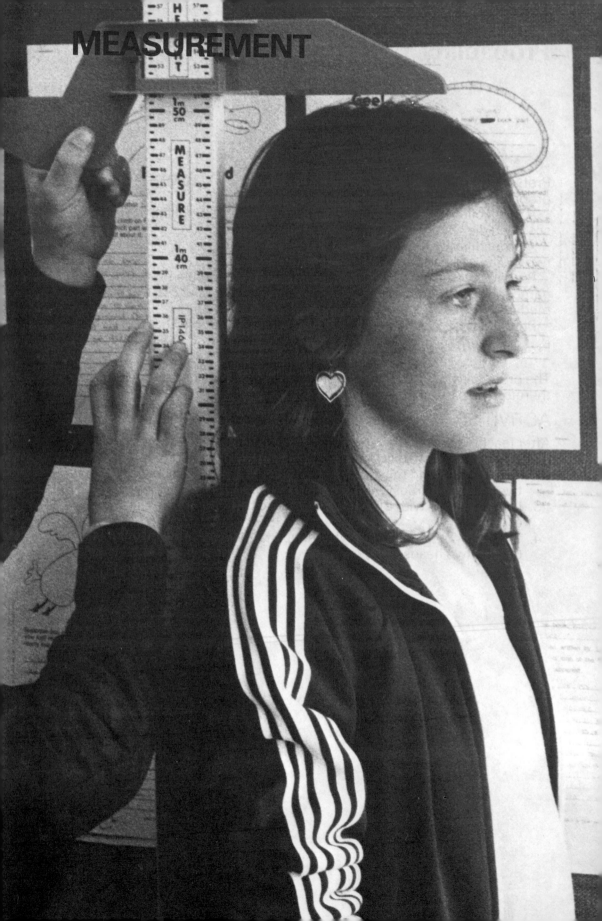

MEASUREMENT

The Measurement Curriculum Guides prepared by the Primary Mathematics Project Team of the Victorian Education Department are used as a basis for the teaching of the nine measurement topics.

To avoid unnecessary duplication of activities, the flow charts (reproduced on the following pages) at the beginning of the measurement booklets have been developed according to grade levels. This sequence of development is a result of teachers using the suggested activities over a two-year period and then reaching agreement on grade level expectations.

Although the aim of the mathematics course states that children should proceed through an ungraded course, experience has shown that children of differing abilities in pure mathematics can all benefit from working through the same measurement activities.

Organizational Suggestions

1. Avoid whole class measurement sessions. There is rarely sufficient material available and the lesson often becomes a teacher demonstration. Children *must have* 'hands-on' experiences.
2. Prepare a measurement activity every week and allow each group to do it on a different day.
3. Find time at the end of the week for all children to discuss the activity together and encourage them to verbalize their developing understandings.

	Length	Perimeter	Area	Mass
Prep.	• Free and directed play • Awareness of the attribute of length • Early estimation experiences • Language of equality and inequality • Measuring with mixed objects	• Free and directed play • Awareness of edges and boundaries	• Free and directed play • Awareness of area as an attribute	• Free and directed play • Comparison of two masses by hand • Comparison of two masses — seesaw experiences • Comparison of two masses using pan balance
1	• Measuring with mixed objects • Comparison of lengths • Ordering of lengths • Length relationships	• Awareness of perimeter • Informal comparison of perimeter	• Direct comparison using regular and irregular shapes • Comparison by covering	• Comparison of two masses using pan balance • Other mass relationships • Size/mass relationships
2	• Measuring in a straight line • Making equal lengths through measuring and comparison • Measuring the difference in units • Measuring with similar objects • Measuring with common units	• Informal comparison of perimeter • Common units used to measure perimeter	• Comparison by covering • Development of suitable common unit	• Comparison and ordering of three masses • Measurement of mass using mixed objects
3	• Measuring with common units • Estimating number of units • Tallying • First formal units • Ruling and measuring using formal units	• Common units used to measure perimeter • Formal units used to measure perimeter	• Covering and counting units and part units • Comparison by cutting	• Measurement of mass using common units • Measuring the differences • Introduction of the first formal unit
4	• First formal units • Ruling and measuring using formal units • Need for larger units	• Formal units used to measure perimeter	• Informal introduction to L × W • Comparison using grids • Introduction of the formal unit	• Introduction of the first formal unit • The use of scales • Need for a smaller unit
5	• Need for a smaller unit • Fractional parts	• Area and perimeter relations • Perimeter related to edges of 3-D shapes	• Relationships between the formal units • Relationships with the number system • Area/perimeter relationships	• Need for a smaller unit • Parts of a kilogram • Introduction of the gram
6	• Inter-relationships of formal units • Inter-relationships with the number system • Development of wider ideas	• Perimeter related to edges of 3-D shapes • Diameter and circumference relationships • Perimeter in the environment • Relating perimeter to the number system	• Volume/surface area relationships • Informal investigations of areas of circles • Informal investigations of area/diameter relationships • Border areas • Further activities related to the child's interests	• Introduction of the gram • Need for a larger unit • Inter-relationships between formal units and the number system

Volume	Time	Money	Visual Representat.	Spatial Relations
Free and directed play — fluids & containers Free and directed play — volume Awareness of capacity of containers Filling and packing	• Language • Routine of the day • Ideas of sequence • Awareness of time cycles and patterns	• Free and directed play	• Free and directed play • Early attempts at representing two objects to show differences	• Free and directed play • Awareness of shape • Recognition and naming of simple shapes • Pattern with shapes • Language of location
Filling and packing Direct comparison of two containers	• Names of days — not in order • Names of days — in order • The hour as shown on the clock face • Experiences with measured but unnamed periods of time	• Recognition of coins • Equivalent values 5c, 10c, 20c, 50c • Shopping activities • Money games	• One symbol representing one object • Two column representation using concrete materials • More than two column representation using concrete materials	• Recognition and naming of simple shapes • Pattern with shapes • Language of location
Direct comparison of two containers Capacity of containers using common units	• Weather chart used to depict week • The week as part of the calendar • Names of the months • Months in order • Comparisons of periods of time to develop ideas of rate and speed	• Shopping activities • Money games • Recognition of notes • Payment for services	• More than two column representation using concrete materials • One symbol representing more than one object	• Pattern with shapes • Language of location • Points of reference and routes developed in the school • Awareness of shapes in the environment
Displacement as a measure of volume Comparing more than two containers Blocks as a common unit of volume	• Names of the months • Months in order • Recording of the date — write name of month • Ideas of named intervals of one minute • Time intervals of 5 and 15 minutes • Reading the time to 5 minutes	• Money games • Recognition of notes • Payment for services • Equivalent values of $1 and $2 • Buying and selling activities	• One symbol representing more than one object • Same data represented in different ways • Use of grids	• Symmetry • Analysis and synthesis of shape • Routes to and from school and map-making
Comparing more than two containers Blocks as a common unit of volume Choosing appropriate units to find the capacity of containers & the volume of fluids Making graduated containers	• Recording the date — months written • Numerals for the month • The calendar as the visual representation of a year • Conventional recording of date • Making and using time measurement instruments • Read time to 1 min.	• Buying and selling activities • Inter-relationships between coin and note values	• Use of grids • Refinement to bar graphs • Development of pictograms	• Relationship between 2-D and 3-D shapes • Nets and skeletons of 3-D shapes • Tessellation of shapes • Routes to and from school and map-making • Point of view
The formal unit to measure capacity of containers & the volume of fluids Volume related to height, base area & surface area Fractional parts of a litre	• Conventional recording of date • Making & using time measuring instruments • Read time to 1 minute • Story of time • Synthesis of the hour • The second as a unit of time • Environmental use of time intervals • Relationships between formal units • Arithmetical operations	• Inter-relationships between coin and note values • Inter-relationship with the number system • The story of money	• Development of pictograms • Locating position on a line (single axis) • Locating position using co-ordinates • Pie graphs	• Nets and skeletons of 3-D shapes • Informal comparison of angles • Tessellation of shapes • Point of view • Points on the compass • Sketch plans
Fractional parts of a litre Measuring differences in capacity of containers The formal unit to measure volume Enrichment	• Story of time • Environmental use of time intervals • Relationships between the formal units of measurement • Arithmetical operations • Century A.D., B.C. • Rate and speed • Planetary relationships • 24-hour clock • Daylight saving	• Inter-relationship with the number system • Interest and percentages • The story of money	• Locating position using co-ordinates • Pie graphs • Line graphs • Enrichment	• Tessellation of shapes • Comparison of angles • Shape and design • Points on the compass • Sketch plans • Grids • Scales and map-making

PROBLEM SOLVING

Mathematical problem solving involves using mathematics in new situations. As children learn mathematical concepts and computational skills, they should also be developing problem-solving skills. With guidance and practice students can develop systematic procedures for solving problems.

Prep. children, when using classification materials such as Attribute Blocks, Logic People and 'What's in the Square?', are beginning to solve problems and develop logical thought. This vital groundwork can be developed further in later grades by being broadened to include simple number sentences as children learn the basic operations.

As children move into the middle and upper sections of the school it is important that regular problem-solving activities be provided. To assist teachers, two sample problem-solving sheets are included at the end of this section. 'Get Smart' is for children in Grades 3 and 4 and 'Mission Impuzzible' for children in Grades 5 and 6. How teachers use the sheets is up to the individual, but the following suggestions should be considered.

1. Encourage children to discuss the strategies they use to solve a problem.

2. Remind children to consider the range of problem-solving strategies when attempting a new problem:
   (a) Draw a diagram
   (b) Use concrete materails
   (c) Use your knowledge of mathematics
   (d) Use trial and error
   (e) Look for patterns
   (f) Estimate, then adjust
   (g) Be on the look-out for problems that require lateral thinking (e.g. When do 11 and 3 more equal 2?)

3. Give children ample time to work on the problems. The sheets could be handed out on Monday and solutions discussed on Friday.

4. Attempt the problems yourself before giving them to children.

5. Be prepared for some of your children to be better problem solvers than yourself. Children will attempt problems with more confidence when they know the teacher sometimes cannot solve a problem.

'Real' Problem Solving

Teachers should be on the look-out for real problems that arise that concern the children or school community, e.g. the school may be planning to set up an area with playground equipment. A small group of senior children could be given the assignment of:
1. suggesting what apparatus should be purchased (from catalogue);
2. costing the project;
3. deciding where the playground could be sited;
4. designing the surrounding area;
5. presenting a written report for School Council's consideration.
 This is real problem solving and children soon learn that real problems:
1. often have many solutions;
2. require value judgements;
3. are more open-ended;
4. sometimes don't have a satisfactory solution.

References

Cook, Marcy. *Think About It! Mathematics Problems of the Day*. Creative Publications.
Palmer, Robert. *Think-Matics*. Martin Educational.
Mathematical Association of Western Australia. *100 Mathematical Problems*. Available from M.A.V.
Creative Publications. *Aftermath*, Books 1-4.

Get Smart

1. Put the numbers 1 to 8 in the squares to make the totals shown.

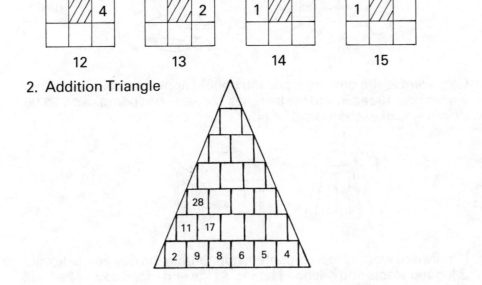

1		
	///	4

12

1		
	///	2

13

	2	
1	///	

14

	4	
1	///	

15

2. Addition Triangle

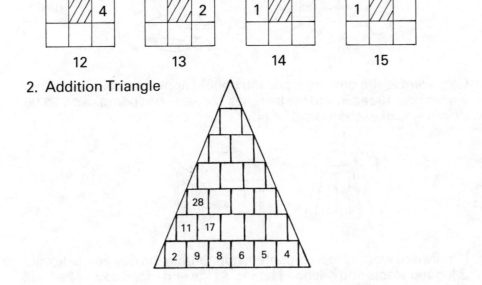

3.

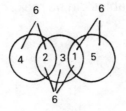

Total 6

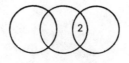

Total 7

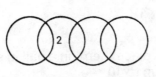

Total 10

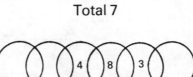

Total 16

Can you make up some more of these on your own?

Mission Impuzzible

1. Gary went to the grocery store and bought apples and bananas. The apples cost 10c each and the bananas 12c each. He spent exactly $1.00. How many of each kind did he buy?

2. Lynn's dad worked in a carpenter shop. During the day he made only 3-legged stools and 4-legged tables. At the end of the day he had used 31 legs. How many stools and how many tables did he make?

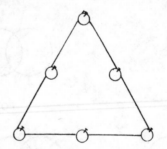

3. Using each of these tags 1 2 3 4 5 6 place them on a nail in such a way that each side of the triangle adds up to 10.

4. I wrote five different counting numbers on five cards. The sum of the numbers is 15. What numbers did I put on the cards?

GAMES AND PUZZLES

Games are an integral part of the mathematics program and this section should be considered in the context of the model for developing mathematical concepts outlined on p.5. Games are not just an extra activity for early finishers or something provided as the occasional special treat. Instead, they are a powerful learning activity for children if used to further develop or reinforce mathematical concepts.

When selecting or making a mathematical game, the following general requirements should be borne in mind:

1. The game should be specifically designed to reinforce a mathematical concept or skill.
2. The game must have an element of skill.
3. The game must have an element of chance or the winner can generally be predicted and many children lose interest.
4. The game must have emotional content. Children need to either compete against other players or contend with the factors of chance or probability.

How to Play

'BUILD A NUMBER'

Materials

1. Sheets for each child with several rectangles divided into three parts.

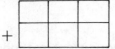

2. Playing cards with digits 0-9.

The teacher or a child turns and reads three cards at random.
The children must place each digit immediately after it is read out and enter it into one of their three-part rectangles.
The teacher can vary the number of places to extend the place-value range.

'BUILD A UNIT'

Materials

1. Mixed Wainwright fraction kit: $\frac{1}{2}$ s, $\frac{1}{3}$ s, $\frac{1}{5}$ s

2. Two dice with one of the fractions $\frac{1}{2}$, $\frac{1}{3}$, $\frac{1}{4}$, $\frac{1}{5}$, $\frac{1}{6}$, $\frac{1}{10}$ on each of the six faces.

Children roll the dice and take the fractional parts shown on the dice from the kit.

Each time a child completes a unit, e.g. $= \frac{1}{3}$ s, a point is scored.

After a set time the child with the most points wins.

'BUILD THE GREATEST SUM'

Materials

1. Sheet for each child with a number of 2 × 3 grids

2. Playing cards with digits 0-9

The teacher or a child shuffles the cards and turns over six cards at random.
Children must place the six digits to make two three digit numbers. When added together, the child with the greatest sum scores a point.

Variations

1. 'Build the Smallest Sum' — Place a decimal point on the chart as shown.

Using the same method as described above, practise the addition of decimals.

2. 'Build the Greatest/Smallest Difference' (subtraction), e.g.

8	7	5
− 4	6	1
4	1	4

3. 'Build the Greatest Product' — Children place four digits, i.e. a three digit number and the number it is multiplied by.

8	6	4
	×	9

7,   7   7   6

'DECICUS'

Materials

1. Three-spike abacus for each player and discs
2. Two coloured dice (the colour is not important) one die = thousandths, the other die = hundredths.

Each player in turn throws the two dice and shows the number they have made on their abacus. The winner of the game is the first person to make one unit.

Variations This game can be used for any place value-range.

'DECI-MAID'

Materials

48 blank playing cards

Make twelve sets of four, e.g.

$\frac{1}{2}$	.5	$\frac{5}{10}$	50%

Do the same for $\frac{1}{4}, \frac{1}{5}, \frac{1}{10}$ etc.

The cards are shuffled and four cards are dealt to each player. The remaining cards are turned face down in the centre with the top card turned up. Each player, in turn, either takes the top turned-up card or the next turned-down card and then discards one card.

The first player to have three of a kind (e.g. $\frac{1}{4}$, .25, 25%) wins.

'EQUIVALENT FRACTION DOMINOES'

Materials

Approximately 25 cards made by the teacher showing equivalent fractions, e.g.

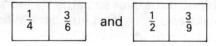

| $\frac{1}{4}$ | $\frac{3}{6}$ | and | $\frac{1}{2}$ | $\frac{3}{9}$ |

Children deal out the cards and the first child to put down all their cards wins.

'EQUIVALENT FRACTION SNAP'

Materials

40-50 blank playing cards with the common fractions written on them —
$\frac{1}{2}, \frac{1}{3}, \frac{1}{4}, \frac{1}{5}, \frac{1}{10}$.

Write equivalent fractions for each fraction on another set of cards, e.g.
$\frac{1}{2} = \frac{2}{4}, \frac{3}{6}, \frac{4}{8}, \frac{5}{10}, \frac{6}{12}$.

Deal cards to all players who, in turn, place their top card face up on the centre pile. When two consecutive cards are equivalent fractions, the first child to say 'snap' takes the cards.

If a child says 'snap' incorrectly, they must give a card to the child on their left. The first child to hold all the cards wins.

'FRACTO' (Equivalent fraction bingo)

Materials

1. Base cards
2. Counters etc.
3. Call cards with common fractions written on them, e.g. $\frac{1}{2}, \frac{1}{3}, \frac{1}{4}$ etc.

$\frac{3}{9}$	$\frac{4}{8}$	$\frac{5}{15}$
$\frac{5}{10}$	$\frac{3}{12}$	$\frac{2}{10}$

The rules are the same as for bingo, i.e. if, for example, 'one half' is called, a child covers $\frac{4}{8}$ and $\frac{5}{10}$ on their base card.

The game ends when a card is covered with counters.

A shorter game can be played where the children only have to cover one full horizontal or vertical row.

'THE HAMBURGLAR GAME'

Materials

S.A. 64 fraction kit.

After children have built a 'Big Mac', e.g. beetroot = $\frac{1}{2}$s layer, gherkin = $\frac{1}{3}$s layer etc., the layers are spread out on the floor.

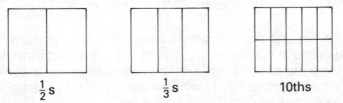

$\frac{1}{2}$s $\qquad$ $\frac{1}{3}$s $\qquad$ 10ths

Each child in turn is the 'Hamburglar' and attempts to steal part of a layer while the rest of the children sit with their eyes closed. Once the bit has been stolen, the children have to open their eyes and say what fraction has been taken. If the missing part can't be identified the Hamburglar has been 'successful' and can keep the stolen piece, but if the missing fraction is named, the part must be returned.

Start the game by only removing one part of a layer, e.g. $\frac{1}{5}$, but later increase the difficulty by stealing more than one part, e.g. $\frac{5}{8}$.

'MAKE A FLAT'

Materials
1. M.A.B. 10
2. Ten-sided die

A group of children roll the die in turn and take from the 'banker' (one of the children) their number in 'minis' (units).

Rule If the children can exchange, they must. If a child receives twelve units, they must exchange ten units for one ten. The first child to make a 'flat' (hundred) wins.

This game can be played with icy-pole sticks. Instead of exchanging, a rubber band is placed around ten icy-pole sticks to form bundles of tens and units.

'BREAK A FLAT'

This game is the reverse of 'Make a Flat' (described above). The children start with a 'flat' (hundred) and give back whatever they roll on the die.

This is an excellent game to reinforce the decomposition method of subtraction.

'MERRY MIX-UP'

Materials
1. Four sets of digits 0-9 written on cards, plus a decimal point card.
2. Acetate boards.

Four teams of two children spread out their cards at the front of the class. The teacher gives a problem and the first team to show the answer with the cards under the acetate board scores a point.

This game can be used to revise most concepts, e.g. place value — 'Show 26 thousandths,$= 2\frac{6}{100}$ or 164 tens'.

'OPERATION CHARADES'

Flashcards with the basic operations written on them, e.g.

$4 + 2 = 6$	$5 - 3 = 2$ take away	$4 - 3 = 1$ difference between
$3 \times 2 = 6$	$8 \div 4 = 2$ shared between	$6 \div 3 = 2$ how many

Children plan how to 'act out' the number sentences and then perform them in front of the grade, e.g. the group acting $2 \times 3 = 6$ would form two groups of three at the front of the grade and then come together to form a group of six. The rest of the class has to write down the number sentence they think has been performed.

Although $2 \times 3 = 6$ would be the preferred answer for this example, $3 + 3 = 6$ would also be correct (inter-relationship of operations).

'PERCENTAGE CONCENTRATION'

Materials
20-30 blank playing cards

The teacher makes 10-15 pairs of cards with matching vulgar fractions and percentages, e.g.

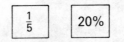

$\frac{1}{5}$ 20%

All the cards are turned down. The children keep matching pairs turned up in front of them.

'ROLL A PRODUCT'

Materials
1. Three ten-sided dice
2. Spinner 2-9

Roll three dice and make the largest number from the three digits e.g. 8 7 4. The spinner provides the number by which the sum is to be multiplied.
   The largest answer (product) scores a point. The child with the most points at the end of a session wins.

'ROLL AN EQUATION'

Materials
1. Two cubes with basic operation signs written on them, e.g. $+$ (2) $-$, $\times$, $\div$
2. Four ten-sided dice

Children in turn roll the six dice and make number sentences. The child with the most number sentences after a set number of turns wins.

'VULCATION'

Materials
1. Playing cards with fractions written on them, e.g. $\frac{1}{2}$ through to $\frac{1}{10}$
2. A ten-sided die

Each child rolls the die and then takes the top card from the shuffled upturned pack and multiplies the two numbers together, e.g.

8      $\frac{1}{5}$ $8 \times \frac{1}{5} = \frac{8}{5}$ or $1\frac{3}{5}$

After every child has had a turn, the children should compare their answers and the child with the largest fraction scores a point. Repeat for a set period.

'WIPE OUT'

Materials
Calculators

The teacher says a three digit number, e.g. 642. The children key in the number on their calculators.
The teacher then gives verbal directions until the number is reduced to zero, e.g. 'Wipe out the digit with the smallest value' or 'Wipe out the digit in the tens place' etc.
   Children who successfully follow the directions and finish with zero score a point.

EVALUATION

Teachers of mathematics are concerned with helping children learn new concepts, skills and knowledge, all of which should involve ongoing, thorough evaluation.

Teachers must be concerned with:
1. each child's rate of progress;
2. each child's level of achievement;
3. whether the learning activities given to children are appropriate;
4. their own role in the prepared learning activities;
5. each child's attitude to the subject;
6. making both subjective and objective assessments of each child's progress so that accurate, meaningful information can be reported to parents and other teachers.

Evaluation involves:
1. determining whether a child has the necessary prerequisites to begin learning a new concept;
2. being satisfied that a child has learnt a new concept;
3. being able to diagnose a child's errors so that the necessary misconceptions can be corrected.

Evaluation Techniques

1. Oral testing — allows the teacher to probe into the child's understanding.
2. Observation — has the advantage of being undertaken during normal working hours over an extended period of time.
3. Pupil records — samples of daily work.
4. Open-ended situations — situations where a variety of approaches are encouraged.
5. Written tests:
   (a) teacher-prepared tests which are valuable in assessing computation, knowledge, understanding, and applications of mathematics.
   (b) published tests, e.g. A.M. Series, CATIM.

Recording Results

The contribution that evaluation makes to the mathematics program will be determined to a large degree by the type and the accuracy of records kept. The difficulty is to organize this information so that strengths and weaknesses of an individual child's mathematical ability are readily revealed.

To enable this type of information to be readily available, 'Mathematics Progress Charts' (available from Audio-Visual Resources Branch of Victorian Education Department) are used in all grade levels. These charts allow teachers space to enter the following information for each child:
1. The section of the course each child is working through;
2. The concepts that have been mastered;
3. Any concepts which require further work;
4. The concept currently being developed.

The following example shows some of the concepts in Section F of the course and the progress of each student. \ indicates that a child has not mastered this concept and ✓ means the child has mastered it.

NAME	PLACE VALUE						
	To 1,000s	10,000s	100,000s	Renaming	Tenths		Equiv. fractions
Mary Green	✓	✓	✓	\			✓
John Brown	✓	✓	✓	\			✓
Bill Grey	✓	✓	✓	\			v
Ann White	✓	✓	✓	\			✓

Measurement concepts can also be recorded on the progress charts to give a clear overview of topics covered, e.g.:

NAME	MASS				VISUAL REPRESENTATION		
	Using common units	Meas. the difference	First formal unit		One symbol rep. more than one object	Same data rep. in many ways	Use of grids
Mary Green	✓	✓	✓		✓	\	
John Brown	✓	✓	✓		✓	\	
Bill Grey	✓	✓	✓		✓	\	
Ann White	✓	✓	✓		✓	\	

In addition to the progress charts, teachers may choose to keep a folder for each child containing examples of class work and written tests and an individual record card with more general comments, e.g. attitude, problem-solving ability, ability to apply known concepts, etc.

The above section evaluations form the basis for the school's end-of-year evaluation program. The shaded-in sections at the different grade levels is the school's end-of-year evaluation program. All children attempt the section evaluations for their grade level, e.g. Grade 5 children do sections E, F and G. The results provide teachers with additional information for end-of-year reporting, as well as measuring the success or otherwise, of the school's mathematics program. The results indicate general trends so that children's performance can be compared from year to year.

End of Year Evaluation Program

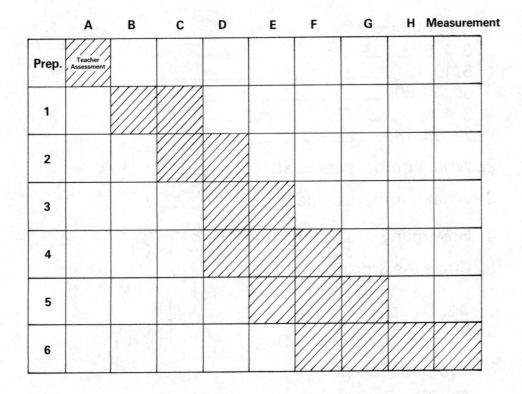

Section B — Evaluation

1. Complete:
   3, 4, 5, _____ , _____ , _____ .
   5, 10, 15, _____ , _____ , _____ .
   30, 40, 50, _____ , _____ , _____ .
   2, 4, 6, _____ , _____ , _____ .
   20, 19, 18, _____ , _____ , _____ .

2. What numbers come after 6 _____ ?

3. What numbers come before 9 _____ ?

4. How many circles are there? 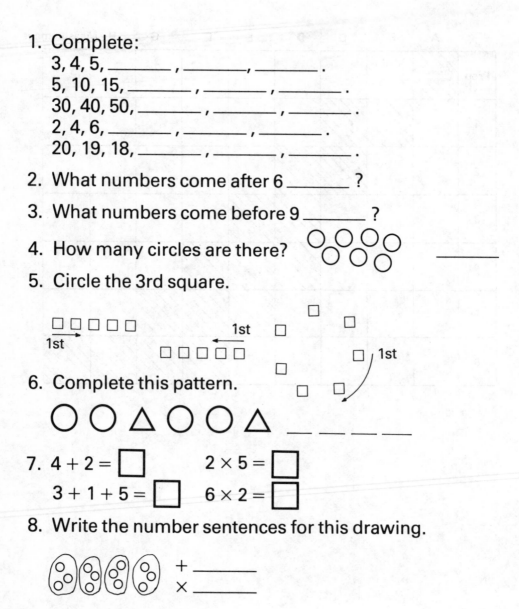   _____

5. Circle the 3rd square.

6. Complete this pattern.

   ○ ○ △ ○ ○ △ __ __ __ __

7. 4 + 2 = ☐   2 × 5 = ☐
   3 + 1 + 5 = ☐   6 × 2 = ☐

8. Write the number sentences for this drawing.

   + _____
   × _____

9. Write the numbers for these words.

two _____   six _____   four _____
eight _____   nine _____   five _____

10. Write the words for these numbers.

1 _____   7 _____
3 _____   10 _____

11. Draw 14 dots in the square.

12. Draw 12 dots in the circle.

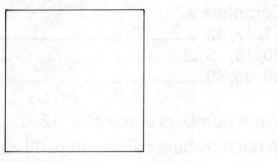

Section C — Evaluation

1. Complete:
   11, 12, 13, _____ , _____ , _____ .
   20, 19, 18, _____ , _____ , _____ .
   30, 35, 40, _____ , _____ , _____ .

2. What numbers come after 12 _____ ?
   What numbers come before 13 _____ ?

3. How many dots are in each box?

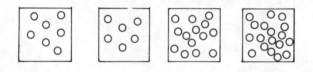

   _____   _____   _____   _____

4. Draw 7 crosses (+) in this shape.

5. Circle the 4th square in these series.

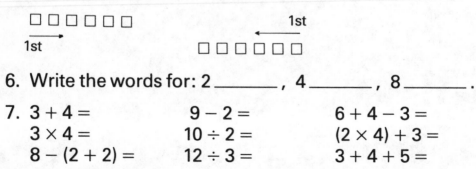

6. Write the words for: 2 _____ , 4 _____ , 8 _____ .

7. $3 + 4 =$ $9 - 2 =$ $6 + 4 - 3 =$
   $3 \times 4 =$ $10 \div 2 =$ $(2 \times 4) + 3 =$
   $8 - (2 + 2) =$ $12 \div 3 =$ $3 + 4 + 5 =$

8. Write the figures for these words.

three _____ six _____ ten _____ five _____

nine _____ four _____ eight _____

two _____

9. Circle the biggest number: 13, 9, 11

10. Circle the smallest number: 26, 62, 39

Section D — Evaluation

1. Complete this pattern.

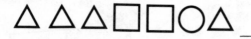 _ _ _ _ _

2. Complete:
   87, 88, 89, _____, _____, _____.
   100, 95, 90, _____, _____, _____.
   31, 41, 51, _____, _____, _____.
   94, 93, 92, _____, _____, _____.

3. What number comes before 70? _____
   What number comes after 89? _____
   What number is two more than 37? _____
   What number is two less than 81? _____

4. Arrange these numbers from smallest to largest.
   36, 103, 4, 63 _____ _____ _____ _____
   Arrange from largest to smallest.
   47, 116, 9, 72 _____ _____ _____ _____

5. Circle the 3rd in each series.

   * * * * * * * *
   →
   1st

   1st
   ←
   □ □ □ □ □ □

6. Double these numbers. 3 _____ 8 _____
   5 _____ 10 _____.
   Halve these numbers. 8 _____ 12 _____
   18 _____ 4 _____.

7. Write the words for 14 _____, 7 _____,
   63 _____.

8. $80 + 6$ = _____ $30 + 9$ = _____
   $7 + 50$ = _____ $4 + 90$ = _____

9. $3 + 4 =$ _____ $3 + 5 + 6 =$ _____
   $3 \times 3 =$ _____ $(4 \times 3) + 5 =$ _____
   $9 - 2 =$ _____ $(2 \times 6) - 8 =$ _____
   $10 \div 2 =$ _____ $(2 \times 4) \times 3 =$ _____

10. $4 + 7$ = $7 +$ _____
    $3 \times 12 = 12 \times$ _____

11. Make four number sentences for 8 using each of the maths signs.
    $(+)$ _____
    $(-)$ _____
    $(\times)$ _____
    $(\div)$ _____

12. Make number sentences for this dot pattern using each of the maths signs.

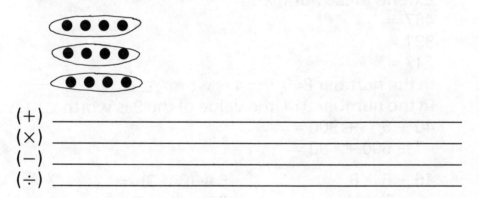

    $(+)$ _____
    $(\times)$ _____
    $(-)$ _____
    $(\div)$ _____

Section E — Evaluation

1. Complete:
   124, 129, 134, _____ _____ _____.
   806, 802, 798, _____ _____ _____.
      4,   8,   16, _____ _____ _____.
   800, 400, 200, _____ _____ _____.

2. $8 + 4 =$ $14 - 9 =$
    $18 + 4 =$ $24 - 9 =$
    $28 + 4 =$ $34 - 9 =$
    $68 + 4 =$ $64 - 9 =$
    $78 + 14 =$ $84 - 19 =$

3. Extend these numbers.
   $467 =$
   $891 =$
   $217 =$
   In the number 846, the 4 is worth _____.
   In the number 914, the value of the 9 is worth _____.
   $40 + 6 \quad + 900 =$
   $7 + 600 + \quad 80 =$

4. $16 + 5 + 8 \quad =$ _____ $6 + (8 \times 3) \quad =$ _____
   $(4 \times 5) + 6 \quad =$ _____ $30 - (3 \times 3) =$ _____
   $20 \div 4 \qquad\quad =$ _____ $4 \times 5 \times 2 \quad\;\; =$ _____
   $30 - 5 - 6 \quad =$ _____ $25 - 6 + 9 \quad\;\; =$ _____

5.   245   362   457
      + 138   + 274   + 268
      —————   —————   —————

          84   42   68
        × 4   × 5   × 3
      —————   —————   —————

6. Bill had 238 marbles and Jane had 682 marbles. How many marbles did they have altogether?

Five boxes each contained 95 oranges. How many oranges were there altogether?

7. What fraction is shaded?

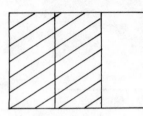 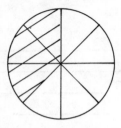 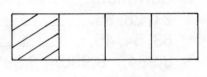

$\frac{1}{2}$ of 6 = $\frac{3}{4}$ of 16 =

$\frac{1}{3}$ of 12 = $\frac{2}{5}$ of 10 =

$\frac{1}{5}$ of 20 = $\frac{3}{8}$ of 24 =

Section F — Evaluation

1. Complete:

   21, 28, 35, _____ _____ _____.

   63, 54, 45, _____ _____ _____.

   8,984,   8,990,   8,996 _____ _____ _____.

   ·6, ·7, ·8, _____, _____, _____.

2.
7 + 5 =	14 − 6 =	3 × 5 =	18 ÷ 9 =
17 + 5 =	24 − 6 =	3 × 50 =	180 ÷ 9 =
27 + 5 =	34 − 6 =	30 × 50 =	180 ÷ 90 =
47 + 5 =	84 − 6 =	300 × 5 = 1,800 ÷ 9 =
67 + 15 =	94 − 16 =	300 × 500 = 1,800 ÷ 900 =

3. Extend these numbers:

   24,386 =

   307,912 =

   2,876 = 28 _____ + 76 _____

   186·4 = _____ tens + _____ tenths

4. Double these numbers:

   30 _____ 48 _____ 243 _____ 370 _____.

   Halve these numbers:

   46 _____ 18 _____ 168 _____ 462 _____.

5.
(3 × 6) + 8 − 6 =	40 − 6 − 3 =
4 × 5 × 2 =	10 − 6 + (3 × 4) =
48 ÷ 4 =	60 − (5 × 10) =
12 + 8 + 6 − 5 =	(5 × 8) − (3 × 6) =

6.

2,614	3,189	684	400
5,384	1,076	− 276	− 163
+ 2,076	+ 5,281		

		164	206
		× 5	× 4

   3) 675   4) 900

7. What fraction is shaded?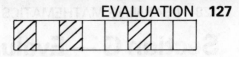

$\frac{1}{6}$ of 30 = $\frac{2}{3}$ of 18 =

$\frac{3}{4}$ of 48 = $\frac{10}{10}$ of 30 =

Supply the missing sign =, > or <

$\frac{1}{5}$ ☐ $\frac{2}{10}$ $\frac{10}{15}$ ☐ $\frac{1}{3}$ $\frac{2}{12}$ ☐ $\frac{2}{24}$

Give three other names for $\frac{1}{4}$

_____ _____ _____ .

8. Work these four problems out on the back of the sheet.
   Show your working.
   (a) Eight children each had 36 marbles. How many
       marbles did they have altogether?
   (b) A school had 346 girls and 298 boys. How many
       children attended the school?
   (c) I wanted to buy a video-recorder. One cost $596
       and the other $836. What was the difference in
       price?
   (d) Five people shared $965 equally among them.
       How much did they each receive?

Section G — Evaluation

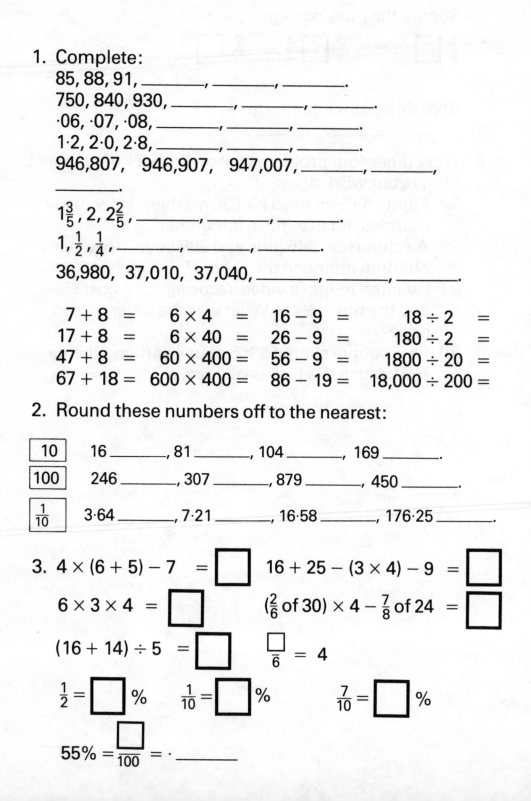

1. Complete:

85, 88, 91, _____, _____, _____.

750, 840, 930, _____, _____, _____.

·06, ·07, ·08, _____, _____, _____.

1·2, 2·0, 2·8, _____, _____, _____.

946,807, 946,907, 947,007, _____, _____,

_____.

$1\frac{3}{5}$, 2, $2\frac{2}{5}$, _____, _____, _____.

1, $\frac{1}{2}$, $\frac{1}{4}$, _____, _____, _____.

36,980, 37,010, 37,040, _____, _____, _____.

7 + 8 =	6 × 4 =	16 − 9 =	18 ÷ 2 =
17 + 8 =	6 × 40 =	26 − 9 =	180 ÷ 2 =
47 + 8 =	60 × 400 =	56 − 9 =	1800 ÷ 20 =
67 + 18 =	600 × 400 =	86 − 19 =	18,000 ÷ 200 =

2. Round these numbers off to the nearest:

10   16 _____, 81 _____, 104 _____, 169 _____.

100   246 _____, 307 _____, 879 _____, 450 _____.

$\frac{1}{10}$ 3·64 _____, 7·21 _____, 16·58 _____, 176·25 _____.

3. $4 \times (6 + 5) - 7 = \square$ $16 + 25 - (3 \times 4) - 9 = \square$

    $6 \times 3 \times 4 = \square$ $(\frac{2}{6} \text{ of } 30) \times 4 - \frac{7}{8} \text{ of } 24 = \square$

    $(16 + 14) \div 5 = \square$ $\frac{\square}{6} = 4$

    $\frac{1}{2} = \square \%$ $\frac{1}{10} = \square \%$ $\frac{7}{10} = \square \%$

    $55\% = \frac{\square}{100} = \cdot\_\_\_\_\_$

4. Extend:

986,421 =

47,238 = 47 _____ + 23 _____ + 8 _____

= 4723 _____ + 8 _____

= 38 _____ + 472 _____

= 47,238 _____

63·49 = 63 _____ + 49 _____

= 9 _____ + 634 _____

= 6 _____ + 349 _____

5.

$$\begin{array}{r} 26,371 \\ 17,806 \\ 9,274 \\ + 10,628 \\ \hline \end{array}$$

$$\begin{array}{r} 6,314 \\ - 2,723 \\ \hline \end{array}$$

$$\begin{array}{r} 2,618 \\ \times 5 \\ \hline \end{array}$$

$$\begin{array}{r} 216 \\ \times 24 \\ \hline \end{array}$$

6) 8574

6. $\frac{2}{6} + \frac{3}{6}$

$\frac{1}{4} + \frac{1}{8}$

$\frac{5}{10} - \frac{2}{10}$

$\frac{2}{3} - \frac{3}{6}$

$4 \times \frac{1}{5}$ $3 \times ·6$

7. Show your working for these below:

(a) 26·12 + 14·87 + 126·09

(b) 28·17 − 16·95

(c) 126·4 × 7

(d) If $2,730 was shared equally between five people, how much would each receive?

(e) If eight bottles each held 740 ml, how much would the eight bottles hold altogether?

(f) What is the difference between 2,647 and 9,600?

Section H — Evaluation

1. Complete:

   1,000,000, 999,994, 999,988, _____, _____.

   ·995, ·996, ·997, _____, _____.

   $2\frac{1}{5}$, $3\frac{2}{5}$, $4\frac{3}{5}$, _____, _____, _____.

   $\frac{1}{8}$, $\frac{1}{4}$, $\frac{3}{8}$, $\frac{1}{2}$, _____, _____, _____.

   1, 7, 2, 14, 3, 21, _____, _____.

   1, 2, 3, 5, 8, 13, 21, 34, _____.

2. $(5 + \square) \times 9 = 108$ $\frac{\square}{5} = 30$

   $4 \times (3 + 8) = \square$ $\square \times 7 = \frac{1}{2}$ of 112

   $\frac{2}{3}$ of $\square = 18$ $284 \times (6 + \square) = 2840$

   $\square \div 4 = 18$ $(7 \times 9) + (9 \times \square) = 17 \times 9$

3. Extend:

   7,648,917 =

   416,814 = 41 _____ + 6,814 _____

   = 416 _____ + 8 _____ + 14 _____

   = 41,681 _____ + 4 _____

   = 81 _____ + 16 _____

   + 4 units + 4 _____

   16,843 = 16 _____ + 843 _____

   = 84 _____ + 16 _____ + 3 _____

   = 168 _____ + 43 _____

   = 16, 843 _____

4. Change to decimals:

   $\frac{7}{10}$ _____, $\frac{16}{100}$ _____,

   $\frac{4}{1000}$ _____, $6\frac{84}{1000}$ _____.

5. Change to vulgar fractions: ·87 _____, ·586 _____,

   4·23 _____, 18·001 _____.

6. Show your working in the spaces provided.

$86,417 + 268 + 9,006 + 28,714 + 9$ 386×79

$42{\cdot}3 \times 0{\cdot}63$ $986,040 - 29,274$

7. What is the probability of rolling a 5 or 6 on a six-sided die?

8. $3^4 =$ _____ $5 \times 5 \times 5 = 5^{\square}$ $32_{\text{four}} = \square_{\text{ten}}$
   $10^5 =$ _____ $2^4 = 2 \times 2 \times 2 \times \square$
   $14_{\text{ten}} = \square_{\text{three}}$

9. $\frac{1}{2} + \frac{1}{3}$ $2\frac{1}{4} + 3\frac{1}{5}$ $4 \div \frac{1}{3}$
   $\frac{2}{3} - \frac{1}{4}$ $\frac{1}{4} \times \frac{1}{5}$ $\frac{1}{2} \div 2$
   $1\frac{1}{4} - \frac{3}{5}$ 8% of $32   25% of 60
   3% of $80 $34{\cdot}2 \div {\cdot}9$

10. Share $1141·56 equally among nine people. How much does each receive?

11. Increase 164 to make it 37 times greater.

12. What is the average of 72, 128, 6, 19 and 35?

13. How much greater is $976 than $242·68?

Measurement Evaluation

LENGTH

1. Measure these lines accurately to the nearest millimetre.

    _____   _____
    A   B

2. 264 cm = _____ metres

3. 186 m × 9 = _____ kilometres

4. If you have run 736 metres, how much further must you go to run one kilometre?

PERIMETER AND AREA

Find the perimeter and area of each shape.

5. perimeter = _____   6. perimeter = _____
   area = _____   area = _____

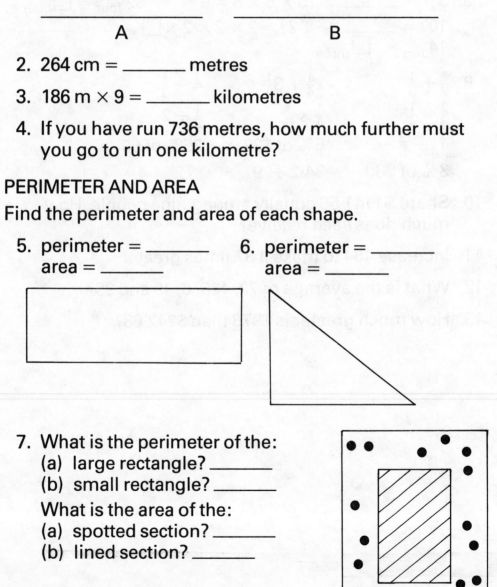

7. What is the perimeter of the:
   (a) large rectangle? _____
   (b) small rectangle? _____

   What is the area of the:
   (a) spotted section? _____
   (b) lined section? _____

VOLUME AND CAPACITY

8. What is the volume of this block? _____

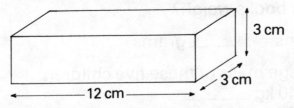

3 cm

3 cm

12 cm

9. How much water would a container hold with a length of 8 cm, a width of 5 cm and a height of 4 cm? _____ The container is a rectangular prism.

10. If I drink 82 ml of milk from a litre carton, how much is left?

11.

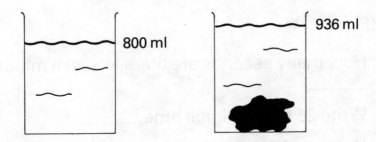

800 ml

936 ml

This is the same container of water after a rock has been submerged in it.
What is the volume of the rock? _____

12. How many centimetre cubes are needed to make a cube of three-centimetre sides? _____

13. How many days would a 150 ml bottle of medicine last if you had to take 5 ml three times a day?

14.

This block has a volume of 72cm³. What could be its:
length? _____
width? _____
height? _____

(Block not drawn to actual size.)

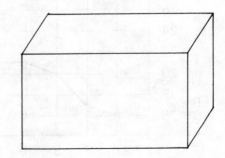

MASS

15. If one exercise book has a mass of 220 g, what would eight exercise books weigh?

16. 3·472 kilograms = _____ grams

17. Find the average mass of these five children:
    Jane weighs 40 kg _____
    Mary weighs 32 kg _____
    Bob weighs 36 kg _____
    Lisa weighs 30 kg _____
    John weighs 37 kg _____

18. If one litre of water weighs one kilogram, what would 250 ml of water weigh? _____

TIME

19. How many seconds are there in seven minutes?
    _____

20. Write 25 to 9 in digital time. _____

21. How much longer than thirty-five minutes is $1\frac{1}{2}$ hours?

22. An aeroplane left Sydney at 11:58 a.m. and reached Hobart at 2:17 p.m. How long did the flight take?

VISUAL REPRESENTATION (GRAPHS)

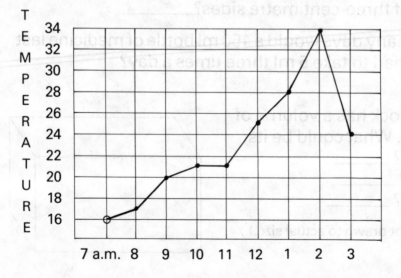

23. When did the temperature reach its peak? _____

24. At what time did the cool change come? _____

25. Between which times did the temperature rise the most? _____

26. What was the temperature at 10.30 a.m.?

CO-ORDINATES

27. Draw the shape made by joining the points:
    (1,2); (3,4); (5, 4); (7, 2)

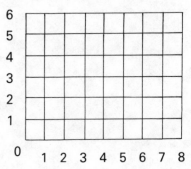

SPATIAL RELATIONS

28. Measure these angles.

    _____   _____   _____

29. Put a tick in the shapes that will tessellate.

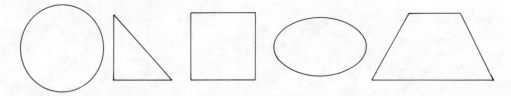

30. What three-dimensional shape will this plan fold in to?

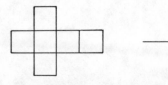 _____

PROGRAM BUDGETING

There are many trial schools in Victoria currently introducing program budgeting. It is anticipated that all schools will be required to adopt this approach, thus enabling effective decision-making and resource allocation within the framework of each school's policies.

All school activities will need to be described in terms of program budgets which state:
1. program goals
2. program description
3. program implementation
4. program resources
5. program evaluation.

Those responsible for writing mathematics programs, be they mathematics co-ordinators, curriculum committees or senior staff, may find the following program budget written for the 1985 school year a useful model.

Schools will already have their program goals stated in their mathematics curriculum documents, but will need to carefully examine current implementation practices across the school to determine whether the program description and implementation is consistent with the expressed goals.

The program evaluation should consider the appropriateness of the program, teacher performance and children's progress and achievement. It is recommended that the program budget statement for a single program not exceed two pages.

BORONIA HEIGHTS PRIMARY SCHOOL

PROGRAM ___Mathematics P-6___

RESPONSIBILITY ___Ross McKeown___ CODE _____

1. PROGRAM GOALS

All students will:

(a) progress through the ungraded course at a rate commensurate with their ability.

(b) learn through understanding by the manipulation of concrete materials.

(c) be given every opportunity to develop excellent problem-solving strategies.

(d) be catered for within their own classroom at their individual level.

2. PROGRAM DESCRIPTION

(a) In each grade students are grouped according to their ability and the organizational requirements of the teachers.

(b) Students commence Section A of our course in Prep. and proceed through the program for seven years. It is anticipated that by the end of seven years most children will have at least mastered Section G.

(c) The students work through the nine measurement topics over seven years. The planned sequence of these follows grade levels, with each topic being divided into seven units of work.

3. PROGRAM IMPLEMENTATION

(a) A mathematics resource room houses all the school's maths materials. A borrowing system operates for all staff members. New materials are added as needs become evident.

(b) The maths co-ordinator teaches across the school as requested by classroom teachers, assists with preparation and plans programs and activities. On-going in-service is provided wherever there is staff interest or need.

  The co-ordinator also carries out individual diagnosis of students identified by classroom teachers.

(c) Monthly planning meetings are carried out at grade levels. These meetings evaluate work covered, and decide on new concepts to be learnt by the various ability groups.

(d) Weekly problem-solving sheets are provided for students in Grades 3-6.

(e) Teacher reference books are being constantly added to the library to enrich the quality of planned learning activities.

(f) Concept development sheets are provided for staff members for the range of concepts covered.

4. PROGRAM RESOURCES

(a)	Paper for problem solving sheets for Grades 3-6 35 weeks × 1 ream @ $3·50	$ 122·50
(b)	Teacher reference books	$ 200·00
(c)	Mathematics materials (pure maths)	$ 400·00
(d)	Mathematics materials (measurement)	$ 300·00
(e)	Laminating of activity cards and mathematical games	$ 50·00
(f)	Mathematical posters, pictures etc. for school, classroom displays	$ 50·00
	Total	$1,122·50

Personnel

Curriculum development officer (Mathematics)
Band 3. $ 30,873·00

Classroom Teachers — mathematics program
24 classrooms, 1 hour per day, 5 days per week
40 weeks, $20 per hour. $ 96,000·00

$126,873·00

5. PROGRAM EVALUATION

Mathematics progress charts are kept by all classroom teachers.
These charts provide the following information:
(a) ability groups in grade;
(b) concepts mastered;
(c) an overview of progress made throughout the year.
This information is gained by the following evaluation
techniques:
(a) oral testing
(b) observation
(c) pupil records
(d) school devised criterion referenced tests.
These charts only *complement* teachers' records.
   A school evaluation program is carried out each year in late
November using school-devised section evaluations. These
results are used to evaluate the effectiveness of the overall
program.